DISEASES OF PLANTS.

DISEASES OF PLANTS

H. MARSHALL

AKASHDEEP PUBLISHING HOUSE
NEW DELHI-110 002 (INDIA)

AKASHDEEP PUBLISHING HOUSE
4374/4B, Ansari Road, Daryaganj
New Delhi-110 002

Diseases of Plants

First Reprint 1993
Reprint 1998
ISBN 81-7158-246-X

PRINTED IN INDIA

Published by Akashdeep Publishing House, New Delhi-110 002 and Printed at Taruna Offset Press, Delhi.

CONTENTS.

DISEASES OF PLANTS

CHAPTER I.

INTRODUCTORY.

DISEASES of plants have attracted attention from very early times, simply because man has cultivated many kinds of plants from of old, and the vines, figs, cereals, olives, &c. &c., were not always healthy in the days of the ancients, as is sufficiently attested by early references to the blasting and mildewing of corn, the casting off and stinking of grapes, bad figs, &c.; but it was long before these maladies were regarded as diseases due to ascertainable causes, and not as evidence of the displeasure of the deities. In no region of science has superstition held sway more tenaciously, perhaps; and even to-day there are not wanting country people who believe that "blights" and "mildews" are mysterious visitations, into the causation of which it is hopeless or even impious to pry, while hundreds of better informed persons will talk vaguely of "electric disturbances" and "atmospheric influences" as causes of diseases, oblivious to the fact that the mere substitution of one vague phrase for another is no step towards explaining a phenomenon.

In the eighteenth century a number of observations on particular diseases were published, and early in the present century the first attempts were made to classify and arrange a knowledge of the diseases of plants; but very little progress was apparent for many years, and even in 1833 Turpin and Unger, who knew a good deal about the symptoms of the diseases of plants, thought that many fungi were produced by the transformation of the cell-contents of the plants they grow on.

As time passed, however, the researches of Meyen, Kühn, Berkeley, and others, led to clearer ideas on the matter, a process which was aided by the rapid advances then being made in the study of the physiology of plants, and in 1866 the foundations of our modern knowledge of fungoid diseases and infection were laid by the late Professor De Bary.

It is not necessary to enumerate all the steps which led to the gradual construction of the doctrine now understood, but it will suffice to say that between 1866 and 1889 the amount of research in this department has been enormous, and the literature of the subject has become overwhelming. The most significant fact, however, is the impulse that has been given to agriculture and horticulture, and the time is rapidly approaching when a farmer or a gardener will as little dare to neglect the study of the physiology and pathology of plants, as a surgeon dare practise without a knowledge of anatomy, or a sailor hope to become a captain without studying navigation. Moreover, these are not studies which will bear trifling with, and he who hopes to understand them must take the necessary trouble to learn how to trace the connection between cause and effect, and scientifically reason from the simple to the complex. In this department the days of empiricism are indeed numbered.

DISEASE, ITS MEANING AND KINDS.

It seems at first sight an easy matter to explain what we mean by disease, and yet most readers will probably agree generally with the following statements: Ill-health may be a very real malady to the being who suffers it, and yet none other be prepared to call the sufferer diseased; and such a remark applies to living beings of all kinds, from oxen to mice, and from trees to mosses —they may be in a condition so dangerous and threatening to their own existence, that the least observant would agree to their being called diseased as soon as the fact was demonstrated to him, though otherwise he might go on never so much as suspecting that their health was affected.

The explanation of this puzzle is not difficult: it depends almost entirely on the point of view of the judge. If we examine a fat prize ox, fed up for Christmas fare and stalled for show, the conviction that this unwieldy and helpless animal is in a condition of disease, does not force itself upon us unless we try to imagine ourselves in a similar predicament; but a wild, or even an ordinary domesticated cow or bull, would no doubt decide that the prize ox was in a very parlous state indeed, and so he is from that point of view.

Again, suppose, for the sake of argument, that the condition of a forced rhubarb plant, with its long pink leaf-stalks, and small crumpled yellow leaves, could be submitted to the consideration of the tough robust plants, coming up more slowly in the neighbouring garden, can any one doubt that the plants would vote it the subject of a dire disease? Only the gardener and the master, and others of the alien beings, would object; but then, you see, their point of view is so different—their object is to eat and digest the rhubarb.

So it is with Strassburg geese and stoneless grapes; from the point of view of the goose or the grape, so to speak, there is organic disease—in the one case an abnormal state of the liver, which prevents that organ from doing its duties as usual; in the other case an inability to produce seeds for the continuance of the species.

It needs no expert to show that multitudes of such cases as these exist around us, and the conviction arises that all beings which are in such abnormal conditions are only able to exist because they are taken care of by man; in other words, they are cultivated or domesticated. Moreover, if we turned them adrift from the watchful attentions of their human masters, they would die; that is to say, their present condition is one that threatens their independent existence as animals and plants. Strictly considered this applies to very many cases which are by no means obvious, and, confining our attention to plants, I need only mention as examples, bleached celery, asparagus, and endive, &c.; double stocks, hollyhocks, dahlias, and other florists' flowers;[1] seedless fruits of various kinds, and so on.

More scientifically put, disease is a condition in which the functions of the organism are improperly discharged; or, in other words, it is a state which is physiologically abnormal and threatens the life of the being, or organ, unless special means are taken to preserve it. And this applies to plants as truly as to animals.

But for our purposes it will be well to adopt a less stringent definition of the idea of disease, because we have to concede something to the customary views of the gardener, farmer, forester, and others to whom these remarks are especially addressed.

We will, therefore, regard as disease only those disturbances of the structure and functions of the plant which actually threaten the life of the plants, or at least their existence as useful objects of

[1] Only quite recently, the interesting discovery has been made that certain gall-forming parasitic mites induce the doubling of flowers by their injuries: the discoverer states that he has succeeded in causing flowers to become double by infecting them with these mites.

culture, as judged by the cultivator whose pocket is concerned.

In any case of a disease there are two factors of first importance to be considered: first, the cause of the diseased condition—the shock or impulse from outside which disturbs the normal condition of the plant; and, secondly, the state of the plant itself at the time. We might put this shortly as, first, the cause of the disease, and secondly, the condition of the patient. For instance, a short time ago I had an opportunity of seeing some ash-plants in a wood, some of which were always suffering from a well-known form of disease, while others were never attacked by it: it was proved that frost caused the injury, and it turned out that all the diseased plants were on a low-lying piece of soil, and differed from their more healthy companions chiefly in being more "watery" at the time when the frosts came. The *cause* of the diseased condition was frost, *i.e.* a temperature far below the normal; but the watery *state* of the shoots had to be regarded also, as rendering the tissues less resistent to the damage done.

Again, certain parasites can attack and kill a given plant when it is young, and all its tissues are soft and moist; but the same parasites may be unable to injure it at all when it grows older.

These are only two examples selected from very many of which we have information; but there are still numerous cases concerning which the cause of the disease, or the condition of the patient, or both, are as yet unknown, and accordingly little can as yet be said or done regarding them.

Even the above necessarily brief remarks will have made it clear that in order to understand the diseases of plants, the inquirer must know something about plants in a healthy state; he should know how they absorb and assimilate their food-materials, how they respire and grow, and so on.

But this is only another way of saying that he should be acquainted with the elements of the physiology of plants, a study which pre-supposes some knowledge of their structure, and the dispositions and relations of their parts. We should think very poorly of any person who undertook to examine and report upon or repair a watch or a steam-engine, if we discovered that he was ignorant of the proper structure and arrangement of the watch or steam-engine, and that he did not know how they ought to work. It is for the same reason that we wisely endeavour to make sure that our medical men are properly instructed in anatomy and animal physiology before we trust them to set us to rights when diseases attack us and throw our works out of order.

Nevertheless, notwithstanding the fact that he who is to really comprehend the diseases of plants, must be well acquainted with their normal anatomy and physiology, it is still true that some interesting and instructive information can be gained about the common and best-known maladies, and especially about their causes, without it being necessary to toil at all the preliminary details; but it is nowhere more prominent than in this branch of knowledge that accurate information of many kinds is likely to turn out useful.

My object, therefore, will be to put forward some facts about the diseases of plants, showing as far as opportunity admits how they mar the normal working of the victim, and especially how they cause it to become less valuable than it would have been in its normal state. Moreover, this volume will be devoted to the consideration of diseases caused by parasitic fungi, as being at once among the most interesting and important of the farmer's foes, and the enemies of gardens, fields, and forests.

In a complete treatise on the diseases of plants, it would be necessary to decide on some plan of classifying the various maladies, according to the symptoms shown by the victim, or the causes which produce the symptoms; but since the present work is confined to the study of certain classes of pests, I shall say no more on this subject just now than the following.

The diseases of plants are caused by two chief series of influences: (1) the injurious action of the external, non-living world, such as unfavourable soil, air, poisons, temperature, &c., or (2) the attacks of living foes, especially parasitic insects and fungi. However, as we shall see further on, it is only rarely the case that one alone of these causes is responsible for the whole of the mischief, for experience shows that many conditions of the non-living environment act in the double capacity of being directly unfavourable to the plant attacked, and favourable to the foe which attacks it, or conversely. Consequently it is impossible to keep the two classes of causes entirely separate.

Another possible plan of classification of the diseases of plants is to group them according to the symptoms, and speak of *Chlorosis*, *Canker*, *Rust*, *Galls*, &c. &c.; but apart from other drawbacks, this method is far too technical for our purposes. I therefore propose to employ a plan which commits us to none of the above difficulties, so far as we need go, and at the same time makes it easy to call attention to the leading features of each disease as noticed by "practical" men. This plan is simply to group the diseases according to the organs attacked, and the fungi which attack them: thus we have root-diseases, leaf-diseases, fruit-diseases, &c.; diseases of seedlings, diseases of timber, diseases of the cortex, and so forth, each caused by a particular fungus. It is true, this plan could hardly be followed in a complete treatise, because it would involve us in difficulties no less formidable than others; but as this book is to be concerned only with certain fungus-pests, and the diseases which they produce, it will be found quite satisfactory for the purpose, and has the great recommendation of being simple. Further, I propose to so arrange matters that, while selecting typical examples of the more important diseases, the various illustrations chosen shall carry the reader step by step along the chief paths of biological research in this domain. It is of course impossible to give "complete lessons in the science of mycology" within the limits of a small book like the present one, but it is hoped that the sketches or studies chosen will give the more serious reader some rallying points, around which he may group further facts and information.

CHAPTER II.

FUNGI, SAPROPHYTES AND PARASITES.

EVERYONE knows what a "fungus" is, so long as you do not press him to attempt more than a popular description of a mushroom or a toadstool, or some other of the commoner and larger kinds met with in our lanes, fields, and plantations; but when it comes to speaking of the thousands of minute forms totally invisible to the unaided eye, it is perhaps too much to expect that he can give any information even as to their existence. Moreover, the story you are likely to get concerning even a common mushroom or toadstool will very probably amount to no more than remarks on the shape, colour, and size of the parts above ground, though this is by no means all, or even the most interesting part of the knowledge to hand.

Unfortunately, nearly all the fungi which cause diseases are extremely minute, and many of them can never be seen without skilled microscopic examination, and consequently the reader who is unacquainted with such matters will have to accept on trust many statements, the verification of which depends on labour of the most patient and refined nature. By the aid of illustrations, however, and by means of simple language, I hope to convey some ideas on the subject as important as they are interesting.

If any one will take the trouble to observe a piece of half-rotten fruit, or mouldy bread, or jam which has gone bad; or a heap of horse-dung, in damp weather, such as we usually have in September; or better still, if he will place such substances on a dish, covered with a bell-jar, and kept damp for a day or two in a fairly warm place, he is pretty certain to find growing on the decaying mass a dense crop of silky gray fungus, the very fine threads of which stand off from the substratum, much as the pile of a velvet stands upright from the rest of the stuff.

This silky gray fungus (Fig. 1) is a mould which has long been known by the name of *Mucor*, a word which may be regarded as derived from the Greek term for a fungus.[1] If some of this is taken and

[1] *Mucor Mucedo* is a very common species, but there are many others.

carefully separated in a drop of water under the

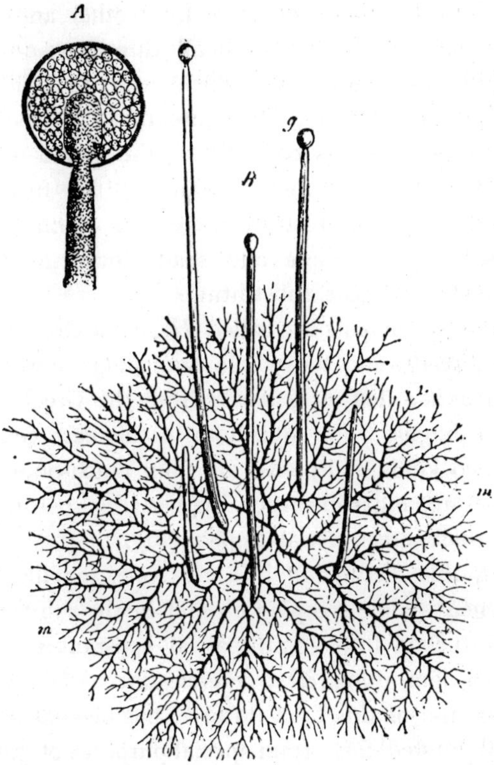

Fig. 1.—A plant of a kind of *Mucor*, showing the mycelium of branched hyphæ at *m*, and the sporangia at *g*. At A is a sporangium much more highly magnified, and showing the spores in its interior. *After Brefeld.*

microscope, the observer can convince himself of the following facts: the fine filaments which stand up into the damp air, arise from other and finer filaments which branch in all directions on and in the substratum, and which serve to hold the fungus in position. There are, then, at least two kinds of filaments constituting the mould—erect, aërial, thicker ones, which stand up free from the substratum, and prostrate or creeping, thinner ones, which support these and send down numerous branches into the substratum.

These filaments are called *Hyphæ*, a Greek word for threads, and the whole branching network of hyphæ is called a *Mycelium*, from the Greek word for a fungus. The mycelium in this case consists of two kinds of hyphæ, (1) thin, much-branched, prostrate, and submerged hyphæ (Fig. 1—*m*), and (2) thick, erect, aërial hyphæ (Fig. 1—*g*). Moreover, this distinction is a real one, for the submerged hyphæ creeping on and in the substratum absorb food from the dissolved substances in the bread, jam, dung, &c.; whereas the aërial hyphæ make use of this food (after it has been absorbed and digested) for certain special purposes of growth and reproduction. Any hypha, more carefully examined, will be found to be a long tube, composed of a thin, transparent, flexible membrane, composed of a substance called cellulose; enclosed in the tube is a glairy, semi-transparent substance called protoplasm, which contains in its mass numbers of drops of water, minute granules, and some other substances (cf. Fig. 7). The protoplasm is the living part of the hyphæ, the cellulose tube is chiefly a protective membrane, and the drops of water (called vacuoles) are employed to dissolve and carry substances from one part to another, and for other purposes of life.

If the above mould is allowed to grow for some days, the tips of the aërial hyphæ will be found to undergo a curious change, each of them swelling up into a tiny ball, like an inflated bladder, only the bladder must be supposed full of protoplasm instead of air. Each of these globular swellings is about as large as a small pin-head, and its protoplasmic contents soon become divided up into some dozens of separate masses, the young *Spores*, from a Greek word for such reproductive bodies. In a short time—the whole process only occupies a few hours when once begun—we find each of the separated masses of protoplasm is itself surrounded with a thin, transparent cellulose membrane, and now the globular swelling may be compared to a bladder full of shot[1] (Fig. 1—A).

The bladder-like swelling on the aërial hypha is termed a *Sporangium*, that is, a spore-case, because it is a case or covering in which the *spores* are produced, as shortly described above. I cannot here dwell upon certain minute points of structure, necessary and important as they are in their proper place, but must briefly summarize what has to be said, directing attention more particularly to certain points.

The globular spore-case gradually becomes brittle, dry, and breaks, scattering the little spores like microscopic shot in all directions.

Now, it is a comparatively easy task to obtain one of these ripe spores immediately on its escape from the sporangium, and if we place it in a drop of water under the microscope, its behaviour can be followed for several days. In a few hours the spore will have swollen, by absorbing water, and under the influence of a suitable temperature, and the presence of the oxygen of the air, it passes

[1] It would require nearly 3000 of these spores side by side to cover one inch, their average size being somewhat less than $\frac{1}{100}$th of a millimeter.

through certain changes known as germination—the spore germinates. At one or two points on its surface the little rounded body puts out a process (cf. Fig. 17 *h*—*k*), the protoplasm bulges out the cellulose membrane, and each process grows longer and longer until it becomes a tubular hypha, with exactly the structure of the hyphæ we examined before. If no further steps are taken these hyphæ will die in a few hours, and nothing more come of our experiment. Why is this? It is because the young hyphæ have exhausted the supplies of food in the spore, and find no food-materials in the drop of water in which we sowed the spore.

If, however, soon after germination has commenced, a drop of sugar-solution, moisture from the horse-dung, or jam, or some other *dead vegetable matter* of the kind, is added, the hyphæ will go on growing longer, will soon branch in all directions, and form a mycelium such as we found at first, and which would then send up aërial hyphæ bearing new sporangia and spores.

Now comes the main point to which I wish to call attention. The above mould, *Mucor*, is a fungus which needs for its existence food-materials derived from the *dead* remains of plants or animals, and since it usually derives its supplies in Nature from their rotting remains, such a fungus is called a *Saprophyte*, from two Greek words meaning a plant which flourishes on rotting remains.

I have chosen a particular species of mould, as above, to illustrate in outline a peculiarity common to very many fungi; they are saprophytes, unable as a rule to injure living plants, but ready to feed upon their remains after death. Like all fungi, these moulds are unable to construct their own carbon compounds (*i. e.* substances such as starch, sugar, cellulose, &c., belonging to a peculiar class of chemical bodies containing carbon) from the carbon-dioxide of the atmosphere; their only mode of obtaining carbon compounds, therefore, is to take them more or less ready made from the remains of plants which contain chlorophyll—the peculiar green substance which gives the well-known colour to all ordinary higher plants—and which are able to manufacture carbon compounds from carbon-dioxide, by the aid of energy derived from the sun's rays.

As I shall have to return to this interesting subject in another place, however, it is sufficient for the present to point out that these mould-fungi are remarkable as living on dead organic remains, and play a great part in Nature by removing them from the face of the earth. Indeed, if it were not for such saprophytes, the gradually accumulating stores of dead organic remains would soon drive all higher living beings out of existence.

But it is evident that such fungi as the above cannot give rise to diseases, since they do not attack living plants.

I will now therefore give a brief description of a fungus allied to the above, which attacks the vine, causing a very bad disease in that valuable plant, and which has cost the vine-growers much annual loss.[1] They know the malady by the name "Mildew," a word which is a corruption of the German *Mehl-thau*, meaning a meal-like deposit on the leaves, and referring to the dusty, floury appearance of the young fungus; but much confusion is produced by the same name being applied to some apparently similar but really very different diseases induced by other fungi. The fungus to be described is called *Peronospora*, from a word referring to the piercing of the leaf by the fungus, and is one of the family to which the dreaded potato-disease fungus belongs.

On the underside of the leaves of the vine there

[1] *Peronospora viticola* is the species here concerned.

are sometimes to be seen pale flecks or patches—(Fig. 2), covered with white, silky, erect, mould-like hyphæ, standing off from the surface of the leaf

Fig. 2—A Vine-leaf attacked by the Mildew (*Peronospora viticola.* De By.) on the lower side. *After Magnus.*

much as the aërial hyphæ of *Mucor* stand off from the substratum (Fig. 3). These aërial hyphæ are traced into the tissues of the leaf, between the cells which they have killed, to a copiously-branched mycelium almost exactly like that of the *Mucor;* the chief difference being that the *Peronospora* mycelium is running between the cells of the vine-leaf—living, dying and dead—instead of between the particles of bread, dung, jam or other such substance (Fig. 4).

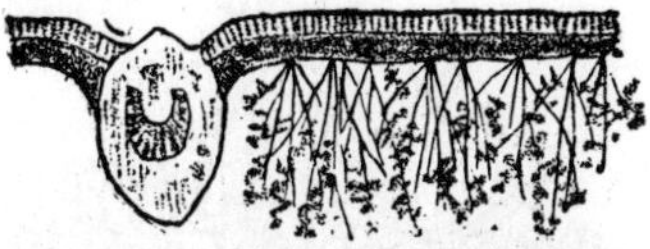

Fig. 3—Part of a transverse section through a Vine-leaf, with numerous tufts of the conidiophores of *Peronospora viticola* standing off from the lower surface. *After Magnus.* Slightly magnified.

Like the *Mucor*, the *Peronospora* bears sporangia, called *Conidia*, from a Greek word referring to the dust-like characters of large quantities of these bodies at the top of the aërial hyphæ, only these conidia are here grouped in compact bunches (Fig. 4); different though the means, the end attained is the same—the conidia become free and scattered at the proper time, and may be collected as were the *Mucor* spores. It is not necessary here to dwell upon the exact resemblances and contrasts between these structures, important as the matter is from a purely scientific point of view. The two features of significance here are the following.

In the first place, the mycelium in the vine-leaf is the part which absorbs and builds up into its structures the food-materials for the rest of the fungus, and which of course holds the latter in position; so far it agrees with the mycelium of *Mucor*.

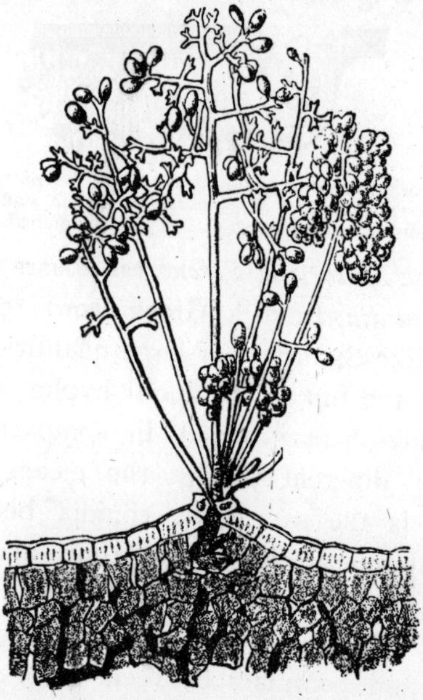

Fig. 4—One of the tufts of conidiophores of Fig. 3, much more highly magnified, together with a portion of the leaf-tissue: the conidia-bearing hyphæ are seen in connection with hyphæ in the leaf, and emerge from a stoma. *After Magnus.*

But the hyphæ of the *Peronospora* mycelium absorb their food-materials *direct from the living cells of the vine-leaf*, piercing these cells by means of minute branches, which act as absorbing organs, and which are termed *Haustoria*, a word meaning suckers. If we now take one of the conidia formed on the aërial hyphæ, and sow it in a drop of water under the microscope, it will swell as did the spore of *Mucor*, absorbing oxygen and water, and if the temperature is suitable it germinates. The details of germination are different from those of the *Mucor* spore, however, for the protoplasm of the conidium breaks up into about a dozen very minute rounded bodies, packed like shot in the cavity of the conidium.[1] Now comes a marvellous change; each of these minute particles of protoplasm begins to move about, at first slowly and then more rapidly, and at last the membrane bursts, and they escape actively into the water. Each of these active little particles[2] is called a *Zoospore*—a word meaning an animal-like spore, and given to them at a time when people supposed that motion was a peculiarly animal character—and high powers of the microscope show that each is an ovoid mass of protoplasm, provided with two

[1] Strictly speaking, these bodies are the homologues of the spores of *Mucor*; the so-called "spore" (conidium) of *Peronospora* corresponding to the sporangium of *Mucor*.

[2] They are so minute that about 2500 would be needed to form a row one inch long.

extremely fine, hair-like threads, which lash the water much as the whip-cord strikes the air when one is cracking a whip; it is the lashing of these active *Cilia* (as they are called, from their hair-like shape), and the reaction of the water, which put the zoospore in movement.

In the drop of water the little zoospores soon come to rest, lose their cilia, and, after undergoing very slight changes, die as did the spores of *Mucor*, simply because no further food-materials are at hand to restore their energy.

Now comes the second feature of significance referred to above. It is of no use to add sugar, or solution of dung, bread, jam, or any *dead* food-materials to the drop of water containing the still living zoospores; they will not germinate the better, or go on living the longer in such substances as did the *Mucor* spores. For the *Peronospora* is a *Parasite*, and not a saprophyte, and is so called because it derives its food-materials from a being which is living. It is thus peculiar even among fungi, in that it not only requires its carbon-compounds ready made, so to speak, by some higher organism, but also in that these and other substances must be presented to it at first hand, as they exist in the living cell.

This may be proved as follows. We have seen that the little zoospores simply die if we try to grow them in dead solutions on a glass slide under the microscope; but if we sow them in a drop of water on a *living* vine-leaf, very different results are obtained. In a few hours the little zoospore will be found to have put out a tubular projection, which bores its way into the vine-leaf, and this germ-tube grows up eventually into a mycelium between the living cells of the leaf, feeding on their contents, and so killing them.

Both in the case of *Mucor*, and in that of *Peronospora*, there are other kinds of reproductive organs formed besides the spores I have mentioned, but reference to these and to several other details is purposely avoided here in view of the special points to which attention is to be directed, and which they do not affect.

These points may be summed up as follows. Fungi are plants which, as they contain no chlorophyll, are unable to manufacture their own carbon-compounds from the carbon-dioxide of the air in the sunlight. Since they cannot exist without such carbonaceous food, however, they are dependent upon other living beings for it, and this in two chief ways. The saprophytic fungi obtain it from the decaying remains of dead plants and animals; the parasitic ones go further, and take it and other compounds forcibly from the plants (or animals) while still alive, robbing their victims of materials which would otherwise have been employed for their own life-purposes. It would almost seem in irony, but the victim is called the *Host:* the meaning of the word is clearer, however, if we have regard to its older significations.

In conclusion, the host suffers injury because the parasite robs it of substance, and incidentally causes disturbances in its economy of various kinds, as will be illustrated as we proceed. But there is every stage to be observed between saprophytes and parasites, from the perfectly harmless mould, to the devastating *Peronospora*, and forms even more destructive; moreover, some fungi are saprophytes during part of their life, and parasites during the remainder, and we have the strongest evidence to show that the one mode of life has led to the other. But, again, and this is a point I shall lay some stress on as we proceed, it is not only the parasite which varies in its habits, and is active towards the host; but the latter also reacts on the parasite, and we shall find that the condition of the host-plant (to keep to plants only) has much to do with the question of parasitism.*

Nay, there are even well-established cases of mutualism, where the host-plant has, so to speak, compromised matters with its foe, and the two—host and parasite—go on living comfortably together, and even helping one another. Of course, in these cases there is no question of disease, as we understand it, and the subject lies beyond our present province.

* A word of warning is necessary here—"the condition of the host-plant" is a phrase employed to express a complex of relations between the plant and its environment, which has been greatly misunderstood. This will be made clearer as we proceed. See Chapter V., for instance.

CHAPTER III.

"DAMPING OFF" OF SEEDLINGS.

WHENEVER mustard and cress are sown very thickly, and kept too damp, a peculiar disease is almost certain to appear, and the young seedlings die off in patches, undergoing a process of rotting of so characteristic a nature that most gardeners know it at once. The seedlings are said to be "damping off," and the disease is one of the commonest of all the maladies found in gardens, occurring in the seed-beds of all kinds of plants in very wet weather, or when the beds are kept too shaded, or where the seeds have been sown too thickly and kept too moist.

The usual course of events is as follows. First one or two of the seedlings are seen to fall over as if top-heavy, and then those immediately around do the same, to be succeeded by others in circles around them, until the whole bed may be destroyed. The plants first affected turn pale, and then become brown and rotten, and then those around follow suit in successive circles, the mischief always starting at the centre and progressing outwards. At last, the centre of the circle of diseased seedlings will be quite rotten, and covered with a white cottony-looking mycelium, very like a mould, and this also will be seen to spread over the circle from the centre to the circumference.

As the disease usually starts from more than one point in a bed, each of these starting-points becomes the centre of a circular patch of dead and dying seedlings, until the edges of contiguous spreading circles overlap, and the patches become confluent. The succession is always the same: the seedlings topple over, turn pale, and gradually blacken and die, and become covered with the cottony mycelium, which weaves them together into matted clumps of dead and dying tissues.

If one of the seedlings which has just toppled over is carefully lifted, and the soil washed from its roots, it will be found that the lower part of the young stem has collapsed, and turned brown or black (Fig. 5—*d*), and if it is placed under the microscope and properly examined, the discolouration will be seen to be due to the destruction of the cells forming the tissues at that spot. Moreover, although no fungus may be visible to the unaided eye, numerous hyphæ, almost exactly like those of *Mucor*, can be seen running between and in the dead cells (Fig. 6), and these hyphæ can be traced

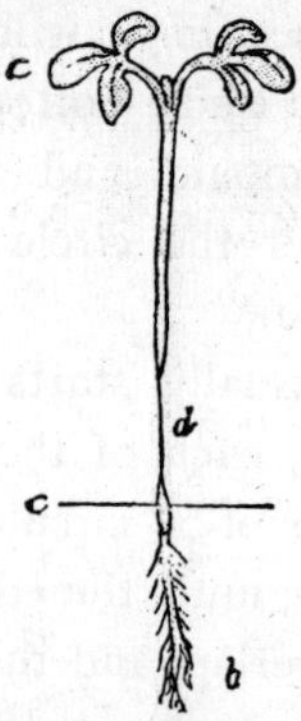

Fig. 5—A young Cress-seedling attacked by *Pythium* at *d*, just above the ground line (*c*) : *b* the root: *a* the cotyledons and plumule. The softer tissues at *d* are quite destroyed, and the seedling topples over at that part. Natural size.

in all directions, stretching up into the stem and down towards the roots, and even coming to the outside and commencing to spread into the soil, or reaching across to another dying cress-seedling.

The seedling has evidently toppled over because the cells at the lower part of the young stem have collapsed and lost their firmness and power of helping to support the upper parts of the seedling. If the selected specimen is kept moist, it will be found that in a few hours the fungus-hyphæ are spreading to the cells above and below the parts already dead, and as they extend to the new cells these also die, one by one, those first touched dying first and later ones in succession. It only takes a few hours—twelve to twenty-

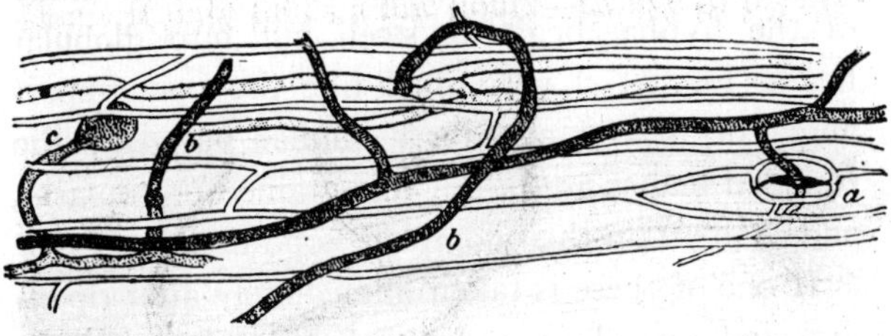

Fig. 6—Portion of the tissues near *d* in Fig. 5 placed under the microscope and highly magnified. The hyphæ of the fungus are seen running in all directions, on and through the cells: at *a* the hypha passes through a stoma: at *c* a young sporangium is about to form.

four perhaps—for the fungus-hyphæ to extend into every part of the seedling, reducing it to a mere putrid mass of blackened cells.

The exciting cause of all this rapid destruction is a fungus allied to the *Mucor* and *Peronospora* already referred to, and called *Pythium*, from a Greek word referring to its capacity for inducing rotting;[1] and the spread of the disease in circular patches is due to the fact that when the fungus has attacked and killed one seedling, it then spreads across to those next it, from these to those around, and so on in ever-widening circles till all are destroyed.

When the *Pythium* has completely destroyed the seedling, or even before this, the ends of many of the hyphæ begin to swell out into globular heads (Fig. 6—*c*), very much like the young sporangia of *Mucor*, and for a similar purpose; the rounded bodies are full of fine-grained protoplasm, and soon attain their complete condition.

If one of these is taken immediately after ripening, and placed in a drop of water on a young living seedling of cress, or other plant which it attacks, or if it is washed or blown naturally on to such a seedling, it soon germinates, provided the temperature is not too low, and that water and oxygen are present. This process of germination is remarkable, because it varies according to circumstances.

In some cases the round body behaves as a spore; it swells, and then puts forth one or more

[1] *Pythium de Baryanum* is the common species, but there are several others.

processes which grow out into short hyphæ on the epidermis of the seedling. If we examine one of

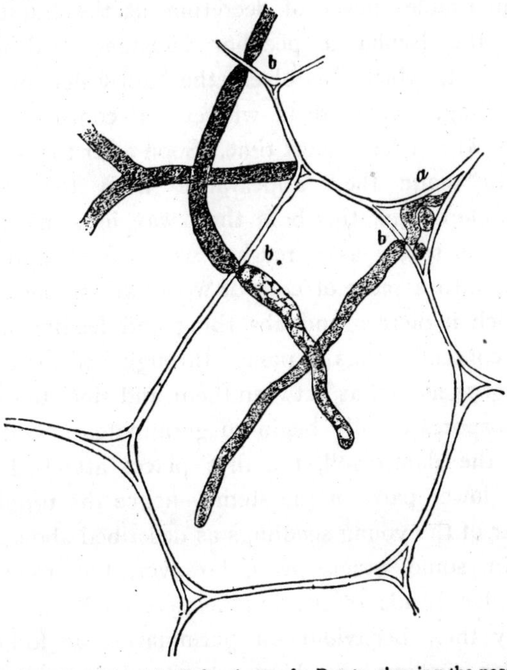

Fig. 7—Small portion of cellular tissue of a Potato, showing the passage of the hyphæ of *Pythium* through the cell-walls at *b*: at *a* are hyphæ in an intercellular space, one of which has then entered the large cell. Very highly magnified.

these hyphæ, (Fig. 7) we find it to be a cylindrical filament with a blunt rounded end, consisting of a tube of thin cellulose containing protoplasm and water, like a hypha of *Mucor* to all appearance. Unlike that mould, however, it has the remarkable power of secreting at the blunt tip of the hypha a peculiar substance called a ferment, which dissolves the cell-walls of the seedling, alive or dead, wherever it comes in contact with it for a short time. Soon after the hyphæ have made their appearance from the spore, therefore, their tips bore their way into the seedling as surely as a red-hot wire would burn its way into a piece of cork or wood, killing each cell which it pierces, and absorbing and feeding upon its contents; thus it passes through cell after cell (Fig. 7), as well as between them, and since the fungus-spores usually begin to germinate, in Nature, on the damp soil, the first places attacked are the lower parts of the stems—hence the toppling over of the young seedlings as described above.

In some other cases, however, the rounded swollen heads of the hyphæ, described on p. 34, vary their behaviour on germination as follows. After swelling, absorbing water and oxygen, &c., as before, the protoplasmic contents drive forwards the cell-wall at a certain spot in the form of a short tube, which looks as if it was going to form a hypha (Fig. 8—*a*); but instead of growing out into a hypha, this short tube swells at the end into a spherical, bladder-like body (Fig. 8—*c*, *d*),

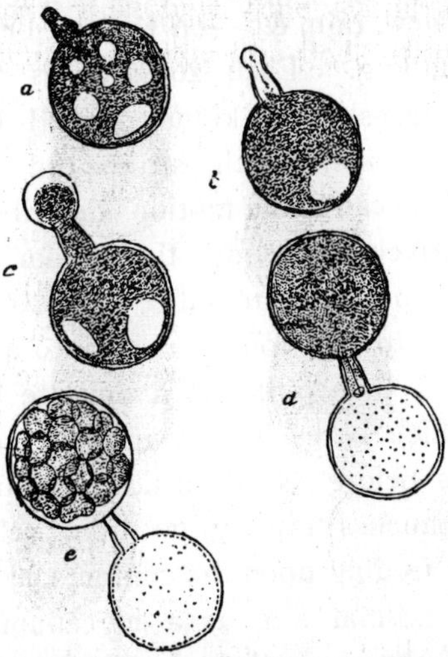

Fig. 8—Germination of a sporangium of *Pythium De Baryanum* in water. The tube put forth at *a* begins to swell at the end, *b* and *c*, and rapidly dilates (*d*) into a vesicle which receives all the protoplasm, which rapidly breaks up into zoospores (*e*). The whole process only occupies about a quarter of an hour. Highly magnified.

into which all the protoplasm collects. The protoplasm then rapidly divides up into about nine or ten minute spheroidal masses, each of which soon begins to writhe and slip about as a tiny zoospore (Fig. 8—*e*); the case in which these zoospores are imprisoned then bursts, and the zoospores escape, swimming actively about in the film of water—dew, rain, &c.—on the soil or epidermis of the seedling. Each zoospore is a minute,[1] colourless, translucent, kidney-shaped body, with two lashing cilia, which strike the water like whips, and so cause the motions described. After moving actively for some time, from a quarter of an hour to twenty minutes, perhaps, each zoospore comes to rest, rounds itself into a ball, loses its cilia, and then puts out a delicate hypha-like process, which bores its way into the cells of the seedling as before. Once inside, the little hypha and its branches rapidly go from cell to cell killing and feeding upon the contents, and growing up to the normal size as a mycelium. In this latter case, therefore, the spheroidal body formed by the hypha (p. 34) acts as a sporangium, which produces zoospores in its interior: in the former case it germinated simply as a spore.

In either case the result is the same, the spore or zoospore produces hyphæ (Fig. 7), which bore into and kill the living cells of the host plant—*i.e.* the seedling. This goes on hour after hour, the fungus developing hundreds of thousands of these spores

[1] The zoospore measures about $\frac{1}{1000}$ of a millimeter.

and zoospores in a few days, and these at once germinate and attack new individuals; whence no wonder need be excited at the marvellously rapid spread of the pest. Moreover, the hyphæ which do not develope spores are able to spread across from one seedling to another, as already said, and the mycelium then becomes so abundant that it can be seen without the microscope, as a white cottony web over the rotting mass.

When matters have come to this pass, it is hopeless to try and save the seedlings; but if one is observant enough to detect the disease in its first stages, when only one or two seedlings are falling over, it is possible to stop the spread of the fungus by carefully removing the affected specimens, and letting in more light and air to the beds; for, note, the fungus is dependent on a plentiful supply of moisture for this rapid development, and the seedlings can usually withstand much drier conditions than the mycelium can.

But now comes a curious fact. It has been found that the seed-beds in which *Pythium* has made its appearance during a given summer and autumn, are extremely apt to be devastated by the same fungus if used again the following spring and summer, and this was explained after the discovery of the phase in the life-history of the fungus, which is about to be described. If one of the dying cress-seedlings, full of mycelium, is kept moist, or better still, put into a tumbler of water for a few days, some of the hyphæ will form spheroidal swellings at their ends, at first so like the spores or sporangia I have described above, that they might easily be mistaken for them, unless they are carefully watched. When one of these bodies has attained its full size, the protoplasm in its interior begins to withdraw itself from the cell-wall, and slowly round itself off into a ball; then a short branch, or the tip of another hypha, grows up and lays its end on the outside, and soon puts forth a short slender tube (Fig. 9), which bores right through the cell-wall and comes in contact with the ball of protoplasm in its interior.

The rounded ball of protoplasm is called an *Oosphere*, a term not easy to translate into popular language, but which means that it is practically an incipient egg, and the globular swelling of the hypha, in which it lies, is called the *Oogonium*—*i.e.* the case in which the egg is contained. The branch, or hypha, above referred to then grows up and lays its tip on the outside of the oogonium ; this branch is called the *Pollinodium*, and the slender tube which arises from this, and penetrates through the oogonium to the oosphere in its interior, is called the *Fertilizing tube*. The pollinodium is full of protoplasm at first, but when the tiny fertilizing tube has reached the oosphere, looking like a slender beak, most of the protoplasm from the pollinodium slowly passes through the fertilizing tube, and is poured gradually into the oosphere (Fig. 9).

After all the protoplasm has passed in that will go, the oosphere becomes surrounded with a delicate membrane, which soon thickens and acquires a tough, elastic character (Fig. 9), and we now find it converted into a perfectly spherical smooth-coated body, which is called an *Oospore*—an egg.

Moreover, and this is the most remarkable point in this curious story, the oospore has acquired some new and very extraordinary properties, owing to the changes it has undergone. The protoplasm that passed into it from the pollinodium has mingled with the protoplasm of the oosphere, and the combination—*i.e.* the protoplasm of the *Oospore*—is now endowed with quite different powers from those proper to it before this union. The process of union or commingling of the two masses of protoplasm is called the process of fertilization; the oospore is the fertilized oosphere.

If these and the following phenomena were not

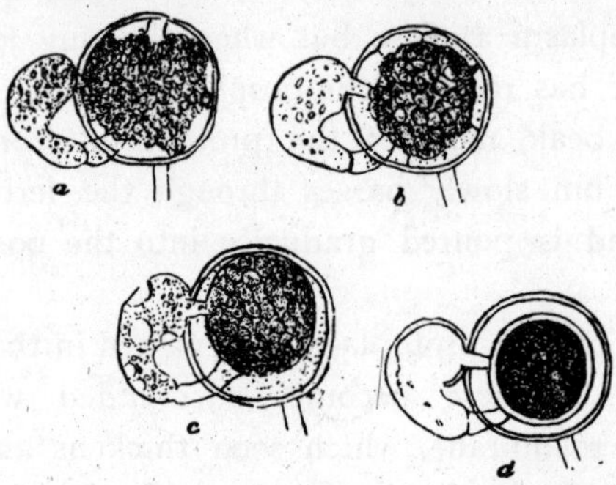

Fig. 9—Development and fertilization of the oosphere of *Pythium*. The granular protoplasm in the oogonium (*a*) is gradually collecting into a ball, and the pollinodium is sending its fertilizing tube into it. In *b* and *c* the fertilizing matter has nearly all passed over into the rounded off oosphere; and in *d* the latter has surrounded itself with its own cell-wall, and now forms the oospore, lying loose in the oosporangium. Very highly magnified.

thoroughly established, having been demonstrated and confirmed by observers of the highest reputation, one might well hesitate before accepting the statements unconditionally; they have been so rigidly investigated, however, that no one need doubt the truth of them, any more than he discredits others of the best results of physical science.

If we take these oospores and sow them in water, as we were supposed to do with the other kinds of spores, they will lie in the water apparently unchanged for days, weeks, even months, and yet not die. It may seem strange that they should not germinate, but they refuse to do so for a long time.

If they are kept over the winter, however, they will germinate in the following spring. Many other cases of the same kind are known in other plants as well as fungi, and all I will say about the matter here is, that the oospore (or rather its protoplasm) requires to lie dormant, sleeping, for some weeks or months before it will re-awaken to activity. It would be beyond my purpose here to speculate as to the kind of changes which the protoplasm undergoes during this long sleep. In any case, they are neither more nor less wonderful than those which the contents of a chrysalis undergo during its period of apparent rest.

Protected in its thick and tough coat, the oospore rests then through the winter, and in the following spring, when water and oxygen are placed at its disposal, and the temperature is favourable, the thick coat bursts, and a hypha is put forth which soon developes spores, zoospores, &c., as before.

Now when one reflects that these oospores are formed in thousands—nay, in bad cases in hundreds of thousands—in the seed-beds during the summer, fall with the rotting seedlings to the ground, and remain at rest during the winter in the soil; there is no longer reason for wonder that the seedlings of the following spring are apt to be badly attacked, if the weather is only sufficiently damp and warm, the seeds sown thickly, and so forth.

One of the reasons for allowing the soil of seed-beds to lie fallow for some time, is that such resting spores may germinate and exhaust themselves in the absence of their favourite food —young seedlings. But, unfortunately for horticulture, the *Pythium* is not wholly dependent on living plants for its supplies of food; it can live for some time as a saprophyte, as well as a parasite, a fact that increases the difficulty of contending with it. In cases where it is very abundant and virulent, the pest may be destroyed by heating the soil of the seed-beds by burning twitch and rubbish on the surface, the great heat produced baking and killing the spores.

I have already stated that cress is not the only plant the seedlings of which are attacked by the *Pythium;* it also devastates sowings of the "Gold of pleasure" (*Camelina sativa*), the "Shepherd's purse" (*Capsella Bursa pastoris*), and some others of the cabbage family. Moreover, its ravages are not confined to plants of this family, for it is found on spurrey (*Spergula arvensis*), on the white Dutch clover (*Trifolium repens*), and will feebly attack even the seedlings of grasses, the tubers of the potato, and some other plants.

There are other seedlings, however, which it will not infect. For instance, all attempts to make it attack those of the flax, poppy, turnip, sainfoin, serradilla, lady's fingers, peas, barley, and some others, have failed.

Consequently it may be inferred that the disease depends in two ways on the condition of the host. In the first place, the host must be of the right species; owing to some specific peculiarities of structure, cell-contents, &c., some plants are able to resist the fungus. In the second place, even a favourable host must be in a certain phase ere the fungus can overcome it. No host is more easily attacked than cress-seedlings, but the younger and feebler the seedling—the more juicy and tender it is—the more easily does it succumb. Old cress plants do not suffer from the disease, and strongly developed, well-aired and lighted, and consequently tough seedlings, may apparently defy it.

CHAPTER IV.

"FINGERS AND TOES," "ANBURY," "CLUB-ROOT."

ONE of the most extraordinary diseases to which plants are subject is that which goes by the names at the head of this chapter, and by other names in different localities. It is particularly abundant among cabbages, turnips, broccoli, cauliflowers, rape, stocks, candytuft, and allied plants. I have found it very common in some parts of Surrey, and it is known all over England, as well as in America and on the Continent, and has been said to reach such alarming proportions that the crops have been all but destroyed in some countries for several years.

On pulling up young cabbage plants for transplanting, the roots are frequently found to be deformed by the development of modular excrescences of all shapes and sizes, from those of a pea to those of a fairly large, irregular, warty turnip (Fig. 10). Such plants are useless, for they develop few and feeble leaves, and all the nutritive matters tend to flow into the deformed roots, making them hard and stringy, when the cultivator wishes them to be soft and succulent, as in the case of turnips. After a time these malformed roots pass into a condition of decay, and the plant—if

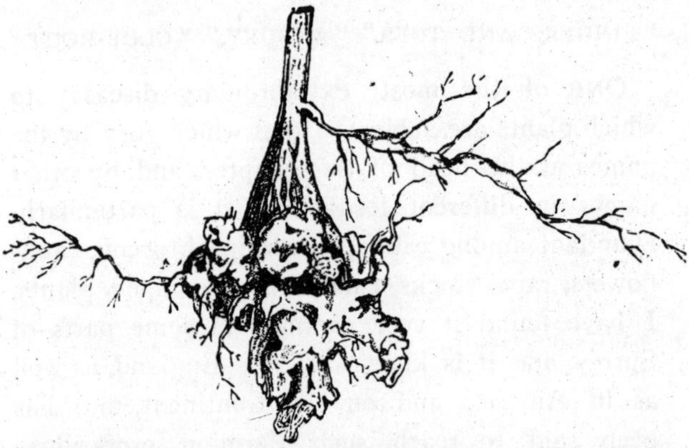

Fig. 10—Portion of a root of a Crucifer, malformed owing to the presence of *Plasmodiophora*. *After Woronin.*

the miserable specimen can be called a plant—dies and rots. If seeds are sown in the same beds, or other young plants of the cabbage tribe are pricked out in the same soil, they are almost certain to suffer from the same disease with even more disastrous effects,

If one of these diseased root-swellings is cut across and examined properly, it is found that the disposition of nearly all the parts is abnormal; the vascular bundles, or strands of woody tissue, are displaced and altered, many of the cells are enormously overgrown, and their contents different from those of healthy cells. The whole structure of the root is interfered with, and its functions disturbed, threatening to destroy the plant—a threat which is often eventually carried into effect most relentlessly.

In the vast majority of cases it is found that all this disaster results from the ravages of a very remarkable organism allied to the fungi, and called *Plasmodiophora*,[1] a name which refers to the slime-like character of the pest: it is, in fact, one of the "Slime-fungi" (the *Myxomycetes*). We may here neglect the rarer cases of somewhat similar swellings on roots due to other causes. The victims are attacked at all ages, but it is especially young plants which suffer.

[1] The particular species is *Plasmodiophora Brassicæ*, because it is a Myxomycete which especially attacks cabbages (*Brassica*) and their allies.

In some of the enormously enlarged cells of the root there may be found a semi-translucent, frothy, and granular slimy mass of protoplasm (Fig. 11), which can be seen to undergo slow movements under the microscope; the granules, drops of water, and other matters in its substance are continually changing their position, and the whole is found to be unquestionably alive. True, the protoplasm in the normal cells is also living, but the more vigorous movements, larger volume, and peculiar frothy nature of this protoplasm in the gigantic cells suffice to show that something out of the common is occurring [illegible] normal cell-contents are, in fact, being devoured and displaced by a curious foreign organism, one of the slime-fungi, as will be seen more clearly as we proceed.

Others of the large cells will be found packed full of minute spherules,[1] and look like bags of small shot (Fig. 12), and the best way of describing what occurs will be to commence with one of these minute spherules, each of which is a kind of spore.

If we isolate one of these spores it appears under a high power of the microscope as a perfectly spherical, smooth, colourless body, clothed by a

[1] The average size is about $\frac{1}{600}$th of a millimeter, *i. e.* more than 15,000 would be needed to cover one inch.

very delicate membrane and containing protoplasm. When I tell you that sixteen thousand of these

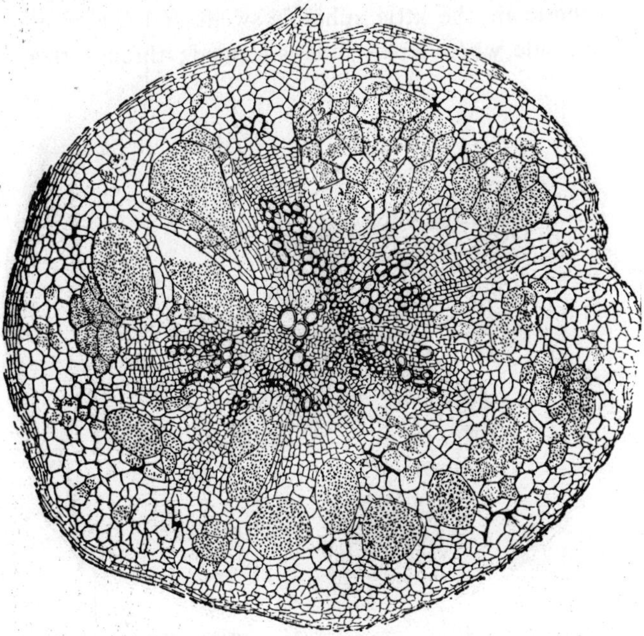

Fig. 11—Highly magnified. Transverse section of a root attacked by *Plasmodiophora*. Numerous giant-cells are seen full of granular protoplasm and young spores: these enlarged cells disturb the arrangement of the other tissues, and act as centres of attraction to food-supplies. *After Woronin.*

spores placed in a row would not cover more than about one inch, some idea of their minuteness may be obtained. After lying for a short time in water, at a proper temperature and in contact with atmospheric air, the little spherule swells and bursts at one side, where a tiny hole is formed; through this

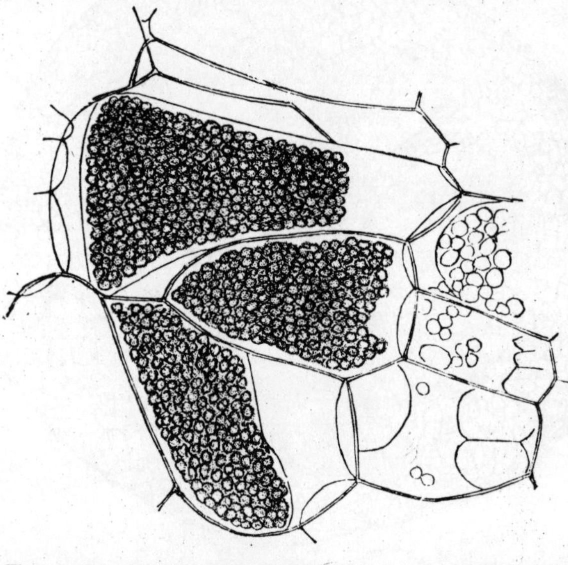

Fig. 12—Several of the giant-cells as in Fig. 11, more highly magnified. Those to the right contain vacuolated plasmodia; the three large ones to the left are filled with masses of spores produced from plasmodia. *After Woronin.*

hole the protoplasmic contents escape (Fig. 13—*b*), and at once begin to move about independently as a writhing, or slowly wriggling little body, called a

Myxamœba (Fig. 13—*d*), a name given to it because it resembles the microscopic organism so common in ponds, ditches, &c., called an *Amœba*. This myxamœba, then, is a mere tiny speck of naked protoplasm, which can move about by writhing or wriggling its substance and changing its form. Small as it is, however, the microscope shows that it has parts. At one spot there is a fine projecting filamentous process, looking as if the glutinous mass had been pulled out into a delicate string; this is a cilium, like the cilia of the zoospores of *Pythium*. Inside the tiny mass of protoplasm a curious phenomenon may be seen: a small, round, clear space appears, grows slowly larger, and then suddenly closes up and disappears—to reappear again, grow for a few seconds, and then disappear suddenly as before—and so on, the phenomenon being repeated every few minutes. The clear round space is a drop of water, a so-called pulsating vacuole, and is produced by the protoplasm pumping water in and out of its substance.

It would require a long chapter to tell of all the peculiarities of this myxamœba; but enough has been said to give some idea of its leading characteristics, and it will be evident that the drops of water in the soil, and the interstices between the bits of rock, &c., which go to make up soil, offer abundance of room for practically any number of such organisms to creep about and live; moreover, all the conditions for their life exist in soil, for the rain-water carries down air in solution, and plenty of dissolved food-materials exist to start the life of the myxamœbæ, if they

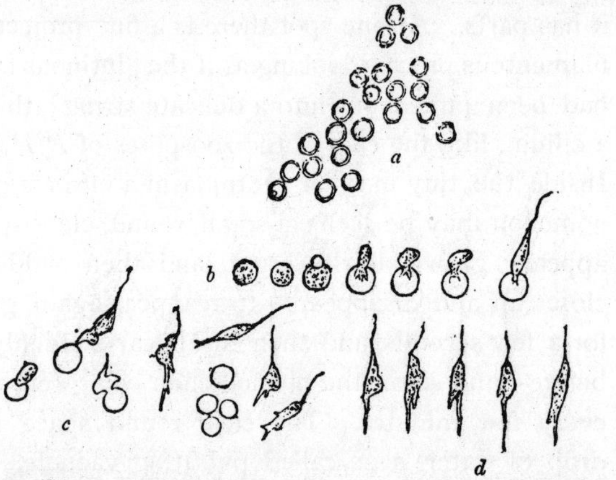

Fig. 13—Spores of *Plasmodiophora Brassicæ*, showing various stages of germination. The spore (*a*) bursts as one point (*b*), and the protoplasmic contents emerge as a ciliated myxamœba (*b*, *c*, *d*), which creeps about by continually changing its form (*d*). Very highly magnified. *After Woronin.*

need them, while the temperature of the soil in spring soon rises to a favourable degree.

For simplicity I supposed that we took one of these tiny spores and watched its germination; but in Nature all the hundreds of thousands of them contained in the malformed roots above described, will be set free in the soil as the roots decay, or into the manure-heaps on which they may be cast to rot. Consequently the soil may become abundantly supplied with the myxamœbæ of *Plasmodiophora.*

If the seeds or young plants (or even older plants for that matter) of cabbages, cauliflowers, turnips, &c., are put into such soil, these myxamœbæ penetrate into the roots, no doubt entering the root-hairs, much as the protoplasm of the zoospores of *Pythium* enters the cells of the cress-seedling, though no one has as yet succeeded in witnessing the entry itself in the former case, as they have in the latter. That the process occurs, however, there can be no doubt, because if we put the plants into soil containing no myxamœbæ, and water them with pure water, they come up sound; whereas if we use the same soil, but scatter some of the spores of the *Plasmodiophora* on it, and water it with water containing the myxamœbæ, all the plants become diseased with "fingers-and-toes" containing the *Plasmodiophora.*

Moreover, it has been shown by experiments that the myxamœbæ can be easily killed by traces of such a poison as carbon-bisulphide, and that if of two beds badly infected with the myxamœbæ of *Plasmodiophora*, one is treated with this substance and the other not, the plants from seeds sown in the one so treated escape, while those in the other suffer from the disease. Of course the seeds are not sown until the carbon-bisulphide has had time to act and to evaporate, as the poison would kill the young seedlings. It is used just long enough to kill the myxamœbæ, and then sowings are made in the disinfected soil.

In cases where the myxamœbæ obtain entry into the cells of the root, they set up a curious action —in fact they commence to battle with the living protoplasm of the cells, both struggling for possession. During this struggle for existence, the abnormally active field of life represented by any particular cell thus attacked, exerts more powerful attraction on the supplies of available food-materials coming from the leaves and soil, and the cell consequently becomes gorged with substance and grows rapidly, and very large. In the end, these increased supplies of food go to the benefit of the conquering myxamœbæ, and these grow up into the large, frothy, moving masses of protoplasm, referred to on p. 50 as occurring in some of the giant-cells (Fig. 11). These large, over-fed masses are no longer called myxamœbæ, but *Plasmodia.*[1] Each plasmodium is, then, a colourless or yellowish, translucent mass of protoplasm, containing water-vacuoles and oily drops, granules, &c., hardly to be distinguished from the normal cell-protoplasm except by its large size and movements, though it really results from a myxamœba which has taken up into itself all the proper protoplasm, as well as other food-materials. But this is not all. The plasmodium is able to pass slowly from one cell to another, devouring their contents, and gaining in size and strength at their expense; and as the area of the struggle extends, its attractive influence on the food-materials prepared in the leaves, also increases in intensity, and consequently the part of the root attacked at length becomes a centre to which all the food-materials flow. The result is that the swellings become larger and larger, and the whole energies of the plant are exhausted in paying heavier and heavier taxes to its relentless foe. Small wonder then that such cabbages, cauliflowers, &c., yield no "hearts" or "heads" for market, for all the food that would normally

[1] This word is derived from the same root as protoplasm, and may be regarded as referring to the protoplasmic nature of the organism.

have gone to make new leaves and flowers, has gone down to the *Plasmodiophora* to try, as it were, to keep the attacked tissues going. In the end this fails—the strain on the plant's energies is more than it can support, and it dies.

Meanwhile, the plasmodia, enlarged and increased in number, as the result of their period of heavy feeding, begin to break up into extremely minute clumps of protoplasm in the cells; each little clump rounds itself off, and is surrounded by a cellulose membrane, and becomes a spore (Fig. 12). The spores lie, as we have seen, in the cells of the root, like shot in a bag, and remain quiescent during the winter, becoming set free into the soil as the root rots, and lying there ready to germinate as before, in the following spring, when their progeny will have good times once more if fortune favours them in the shape of new crops of plants of the cabbage tribe. What they will do if no such plants are put into the ground no man knows, though it is certain that many millions of them die every year. It is not improbable that they can support themselves to some extent as saprophytes, but this is not as yet demonstrated.

CHAPTER V.

THE "POTATO-DISEASE."

It seems impossible to adopt any other name than the above for the disease I am about to describe, though it is by no means the only disease to which potatoes are subject. However, since the malady in question has at various times attained such virulence that large areas of population have been seriously affected by the effects of its ravages, we have weighty reasons for regarding it as *the* potato-disease. Its general history in this country and in the rest of Europe is well known. In 1845-50 the epidemic, which had previously made its appearance in North America, broke out with terrible virulence in this country and on the Continent, and although it has never been so disastrous since, we have never lost it. Although its outward symptoms were more or less familiar during the first half of the present century, the scientific investigation of the fungus which causes the disease was not seriously carried out until the end of that period, and the difficult task was not advanced many stages until a considerably later date.

When the potato plants are in full leaf, in July, it is a common event that some of the leaflets begin to curl over and die, these symptoms being preceded by the appearance of brown patches, which spread from small centres and gradually extend over the leaf (Fig. 14). In very damp, warm, close weather, these brown spots may be seen on nearly every leaf in the garden or field, and they then spread so rapidly that in two or three days the whole field may be "blighted"; every leaf, and even the stalks, being reduced to a blackened, putrefying mass, with a distinctly unpleasant odour.

More carefully observed, the progress of the disease is found to be as follows, in a typical case. Spots, at first pale and then brown, appear on a leaf or leaflet, some at the edges or tips, others in other parts of the leaf-substance, and slowly increase in area and in depth of colour. If the under side of the leaf is examined, a quantity of whitish silky hyphæ may be seen at the margins of the dark-brown spots, and hour after hour, as the dark patch increases in area, the advancing margins are accompanied by the projecting silky hyphæ. If the weather is very warm and moist, and there is no breeze stirring for a day or two; or if the leaf is kept in a perfectly still, damp atmosphere, under

Fig. 14—A potato-leaf, showing the spots and patches of "potato-disease," due to the ravages of *Phytophthora infestans*. In the darker patches the tissues are quite dead: the margins of the spots would show the hyphæ of the fungus, standing off much as in Fig. 3. *After Sorauer.*

a bell-jar, the silky filaments are particularly abundant, and glisten as they reflect the light, and are seen to be accompanied by a very fine whitish powder. These projecting silky filaments are hyphæ of the fungus of the potato-disease, and the glistening whitish powder consists of myriads of its conidia. This fungus has received

Fig. 15—Section of potato-leaf in the tissues of which is the mycelium of *Phytophthora*. The hyphæ run between the cells, and send through the stomata (*a*, *c*, *d*) the aërial branches which bear the conidia, *b*. *f*, one of the peculiar hairs of the underside of the leaf. The dark parts of the tissue of the leaf show where cells are dying from the effects of the parasite. Highly magnified. The normally upper surface of the leaf is here turned downwards.

the name *Phytophthora*,[1] which means a destroyer of plants, and is a not very distant relative of our old friend *Pythium*.

The relations of the fungus to the disease-spot are best made out by good transverse sections through the leaf at the regions above described. It will then be seen that the filaments referred to are aërial hyphæ (Fig. 15), which spring from a mycelium, the branches (hyphæ) of which meander between the cells of the leaf. If such sections are taken through the margin of the disease-spot, some of the cells of the leaf will be seen to be dead, their contents shrivelled, smudgy, and turning brown, and their cell-walls discoloured; whereas others may be found still alive and healthy—their cell-walls sharply defined and colourless, and containing chlorophyll-corpuscles in the protoplasm. Running between the cells, especially where the latter are beginning to die, may be observed

[1] *Phytophthora infestans* (De By.): *Peronospora infestans* (Mont). The fungus has received many other names in the past, but these are the only important ones. It may be noticed here that many plant-diseases are due to parasitic allies of the fungus; for instance, seedlings are killed by a *Phytophthora*, and diseases of the vine, onion, lettuce, beet, spinach, and many other plants, are caused by various species of *Peronospora*.

sinuous tubular hyphæ (Figs. 15 and 16), very like those of *Mucor* or *Pythium*, and which therefore require no detailed description here. Their course lies chiefly along the cavities between the cells—hence called intercellular spaces—and at certain points the ends of the hyphæ may be seen emerging to the exterior, to form the aërial hyphæ referred to above. When they have emerged, and grown out into the damp air necessary for their further development, these aërial hyphæ put forth successive branches bearing conidia. If there is any breeze stirring, these conidia are blown off almost as quickly as they are formed, but if the air is still, the conidia may accumulate in sufficient quantities to give the glistening powdery appearance seen from without. If the air is too dry, the conidia will not be formed, or only very few will develop at intervals, as at night, when the air is temporarily a little moister.

In the healthy green portions of the leaf, at a distance from the disease-spot, no hyphæ will be found, and in order to understand the action of the mycelium as it spreads into these portions, I will here say a few words as to the structure and functions of the healthy leaf, &c. in the potato plant.

The leaf may be looked upon as a thin, flat plate of tissue, consisting of enormous numbers of little bags or cells, most of which are in contact and joined at certain points only, so that passages exist between them (cf. Fig. 15). These bags or cells are composed of thin membranes of cellulose, and each contains a quantity of protoplasm lying on the inside of the membrane, and closely applied to it much as a wall-paper is applied to a wall. This protoplasm is the living cell-substance; the thin, elastic, cellulose membranes being chiefly supports, which, although they practically separate off the protoplasm of one cell from that of another, do not entirely cut off communication between the cells. And this for the following reason (in addition to others which it is beyond my purpose to enlarge upon here): namely, that the cellulose is permeable to protoplasm and water, and food-materials in solution, and is thoroughly saturated with such fluid.

Enclosed in the hollow bag formed by the cellulose and its protoplasmic lining, is a quantity of cell-sap—*i.e.* water in which are dissolved all kinds of substances which have been used or are to be used as food, &c. for the protoplasm. For it must never be forgotten that it is the protoplasm which is the living substance of the leaf, and it requires feeding to keep it alive, just as a myxamœba needs feeding to keep it alive: in fact, the resemblance between the two is much deeper than can be seen from a short sketch like this.

But the protoplasm in each of these cells contains other structures besides, of which I will only mention one here. Scattered in its substance are numerous minute green bodies, looking like brilliant green drops; they are not mere drops, however, but have their own peculiar structure also. These bright green bodies are the chlorophyll-corpuscles, and it is because they are so extremely numerous, in the millions of cells, that the leaf—and the same is true of other green parts of plants—obtains its bright fresh green appearance. The substance which gives the corpuscles this colour is called *Chlorophyll*, a word meaning leaf-green, and it is a substance possessing very remarkable and important properties. Put shortly, whenever the sun shines on one of these green chlorophyll-corpuscles, so long as it is alive and healthy, and has ordinary air and water in contact with it, the corpuscle has the power of doing a very remarkable thing—it seizes and holds fast the small quantities of carbon-dioxide which the air contains,[1] and some of the water, and breaks them up and re-combines their elements, producing the substance which we know as starch, and setting free certain quantities of oxygen gas.

Now a good deal of attention has always been directed to this remarkable phenomenon, for two reasons. In the first place, because the oxygen set free during this process, occurring in the sunlight in every one of the millions of millions of chlorophyll-corpuscles in the green plants around us, is so important a factor to man and other living beings in the economy of nature; for no living organism, plant or animal, can live without oxygen. In the second place, the formation of this starch (and other substances which result from its transformations in plants) means so much food-substance gained by organic beings from inorganic nature; the energy bound up in the starch which has been formed from carbon-dioxide and water, is a real gain from the outside universe —it is power derived from the sun, and held fast until it is made to do work as food or other fuel. If we suppose a man fed entirely on

[1] The atmosphere in the open contains ordinarily about four to six parts of carbon-dioxide in every 10,000 parts of air.

potatoes and pork, the pork being obtained from pigs fed only on potatoes, then the man may be regarded as fed entirely on substances which have been obtained for him from the inorganic world by the potato plant. In the absence of this intermediary, however, he could never use the substances from which his food came—the carbon-dioxide, water, nitrogen, and a few minerals, of which his food consists, are useless to him (for this purpose) as such, and he could not store up the solar rays, as they are here stored for him.

But what I wish the reader to understand is, that important as the matter is from our point of view, this is not quite the proper way to measure or estimate the importance to the plant of this process of assimilation of carbon-dioxide; looked at with relation to the potato, this accumulation of material and energy is solely for the purpose of providing for the needs of the potato itself and its progeny. As fast as starch is made in the green chlorophyll-corpuscles, in every one of the millions of cells of the leaf, it is handed on in a soluble form from cell to cell, down the stalks or haulms, to the young cells which are developing into potato-tubers under the surface of the soil. The formation of the starch in the leaves can only go on during the day, when the sun's rays are shining on the leaves, but this passage onwards of the dissolved substance down the stalks to the tubers, goes on actively in the dark as well. Some of the starch in the leaves is used up on the way; it is employed to make new leaves, buds, flowers, &c., or to increase the size of those already formed, but not fully developed, and a considerable quantity is consumed on the spot. Nevertheless, at the time when the potato plant is in full vigour, in July and August, say, the leaves manufacture far more starch than is required for their present needs; they are storing the surplus in those neat-looking, compact, swollen buds, which we call potato-tubers, where it would normally remain dry and sound till the following spring or summer, when the sprouting tubers would employ it to start their new growth, converting it into new cell-walls, shoots and leaves, &c.

Forcing ourselves into the point of view (so to speak) of the potato plant, it must be allowed that its enemies are numerous and powerful. Its stores of starch—the tubers—acquired with considerable trouble while the sun shone, and the weather was favourable, are prized by many animals, man included, and what is more to our present purpose, by some plants; and by none more than by the fungus of the potato-disease, *Phytophthora infestans.* Not the least astounding fact about this latter, however, is that its direct point of attack is especially the seat of manufacture of the starch, rather than the fully formed starch itself; the fungus attacks the cells of the leaf just at the period when they are most busily and usefully employed in forming and passing on the starch. Herein lies the deadly nature of the pest, that in its epidemic phase it destroys the manufactory, and then follows up its conquest by running down the dying stalks to the stores in the tubers, where it apparently goes to sleep for the winter; of which, however, more will be said presently.

Returning to the normal leaf. Its structure is obviously adapted with reference to its functions; the broad surface of thin leaf-tissue exposes a large area to the sun, light, and air; stiff, elastic strands of fibrous and vascular structures keep it expanded, and allow it to sway in the wind, and at the same time bring water to its cells from the soil, *viâ* the roots and stems. The leaf-tissue proper is covered by a thin, transparent layer of cells, called the epidermis (or skin), which does not prevent light from passing through, although it is sufficiently elastic and tough on the exterior to protect the leaf, and sufficiently impervious to moisture to prevent

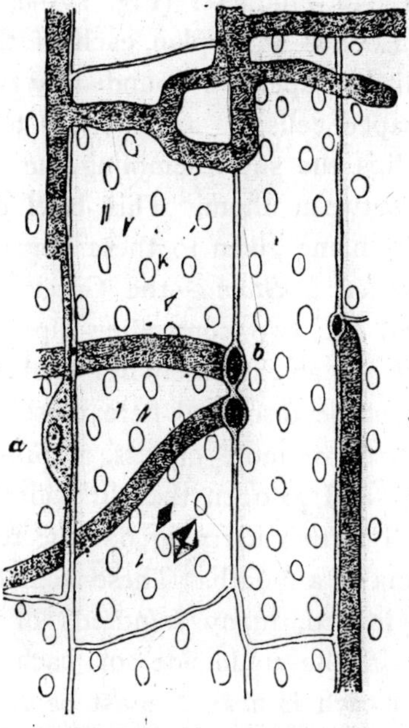

Fig. 16—Piece of tissue of stem of potato-plant, showing the hyphæ of *Phytophthora* running in the cell-walls: (*a*) nucleus of a cell; the other contents shown are crystals and chlorophyll corpuscles. Very highly magnified.

the delicate tissues below from losing all their water by uncontrolled evaporation, and so shrivelling up.

At numerous points on its lower surface, however, the continuity of this impervious epidermis is broken by extremely minute holes or pores of a very singular character (Fig. 15—*a*, *c*, *d*, and Fig. 18). Looked at in plan, each of these pores is a tiny slit-like aperture bounded by two curved, sausage-shaped cells, so arranged as to look like two thick lips, the slit resembling the opening of a mouth between them. This brief description explains the name given to these pores. Each of them is called a *Stoma*—the Greek word for a mouth—and the two controlling lips are called *Guard-cells*. By an extremely beautiful mechanism, which cannot be discussed here, these guard-cells are able to curve more or less, according to circumstances, and so open the slit wider or nearly close it, still further carrying out the resemblance of the stoma to a mouth. These stomata are very numerous indeed, many hundreds of thousands occurring on the underside of each leaf; the aperture of each is also, it must be remembered, extremely small, and only the finest of fine hairs could pass between the guard-cells without injuring them. If now we suppose such a fine hair passed into a stoma, its forward end would lodge in a cavity (Fig. 15—*e*) between some of the cells of the leaf-tissue inside—the delicate cells containing the chlorophyll-corpuscles in such abundance. This cavity between the cells, or intercellular space, as we may call it, communicates with the passages between the cells, referred to on p. 65, whence it follows that if, instead of the hair, we started some minute animalcule in at one of the stomata, it could traverse these passages all over the leaf, if it had sufficient determination and strength to carry out an undertaking of (to it) such enormous magnitude.

Enough has been said to show that air, watery vapour, or other gaseous bodies can penetrate the stomata and traverse these intercellular passages throughout the leaf; and that is just what the gases and vapour of the atmosphere do, and hence the relatively small quantities of carbon-dioxide in the atmosphere are brought close to the chlorophyll-corpuscles, passing in solution through the damp walls of the cells which contain them. Hence, also, the oxygen set free can pass out to the atmosphere, traversing the passages and stomata by diffusion. Hence, moreover, water vapour passes off from the surface of the cells into the passages and air, when the latter is dry; and the replacement of this lost water by other water coming from the roots, which absorb it from the soil, enables the leaf to obtain supplies not only of that valuable fluid, but also of the small quantities of minerals which it brings up in a state of solution. For the same reason, also, the oxygen of the air can reach the cells, without which the protoplasm could not live. In short, all disturbances of the diffusion equilibrium of the gases composing the atmosphere, and those evolved from the cells, tend to be balanced by exchanges through these passages and stomata.

We thus see that the leaf of a potato plant is a very complex and delicate piece of mechanism, and I have by no means described all the intricacies and co-ordinations to be met with in it. But I hope the matter has been made so far clear that the reader will realize how profoundly all these nicely-balanced adaptations may be disturbed if the hyphæ of a fungus, like *Phytophthora*, obtain an entry into these intercellular spaces, and actively set about destroying the unions of the cells and absorbing some of those valuable contents they are so busily manufacturing.

This is just what the hyphæ do accomplish; they make their way between the cells (Figs. 15 and 16), dissolving substances properly belonging to the latter, and then feeding upon the produce.

Now, even if these hyphæ did nothing more than block up the passages between the cells, it is clear from the foregoing that they would prevent the proper aëration of the cells, and the free exchange of gaseous carbon-dioxide, oxygen, and water between the air and the cells; but these hyphæ, like those of all fungi, are very greedy consumers of oxygen, and need much water, and so do further harm by diverting these substances from the leaf-cells. They do most injury, however, by directly absorbing the products of assimilation formed by the chlorophyll-corpuscles, robbing the plant of the materials which would otherwise have been sent down to the tubers, below ground, to be stored up for the winter; and they also rapidly kill the cells and thus cause the patches of dead tissue (Fig. 14) with which we started. When the foliage is destroyed, the mycelium passes down the stalks to the tubers, and if the season is very moist and warm, and the fungus has established its hold early, the hyphæ may at once proceed to destroy the young potatoes. More commonly, however, the tubers are beginning to ripen when the hyphæ reach them, and the mycelium goes to sleep—passes into a dormant state, between the cells of the ripening tubers. It is because the potato-tubers contain these dormant hyphæ which have sneaked in at the last moment, that so many of them go bad in the store-sheds or heaps during the following winter and spring, and such tubers of course if kept dry for "seed" are especially apt to reproduce the disease during the next summer.

I have yet to relate how the fungus spreads so rapidly from plant to plant during a wet summer. If we take one of the ovoid conidia, as they are called, from the aërial hyphæ (Fig. 15—*b*) which spring from the disease-spot (p. 61), and sow it in water, it rapidly undergoes the following changes if the conditions are kept favourable. The narrower free end of the conidium bursts, and emits about ten little zoospores (Fig. 17), very like those of *Pythium*, and developed in the same way from the protoplasmic contents of the conidium. Each of these zoospores flits about actively in the water, by means of its two cilia, for about a quarter of an hour or longer, and then comes to rest, loses the cilia, and rounds itself off into a tiny sphere[1] of protoplasm clothed by a mere film-like membrane of cellulose (Fig. 17, *e*—*k*). It then behaves like a zoospore of *Pythium*, putting forth a very delicate

[1] In size they are not unlike the zoospores already mentioned—*i. e.* about $\frac{1}{2500}$th of an inch in diameter.

hypha, which rapidly grows out as a tube (Fig. 17 *h—k*). In water only, and on glass, no further growth occurs and the young fungus dies; but if the above germination has occurred on a leaf of the potato plant, in a drop of rain-water or dew,

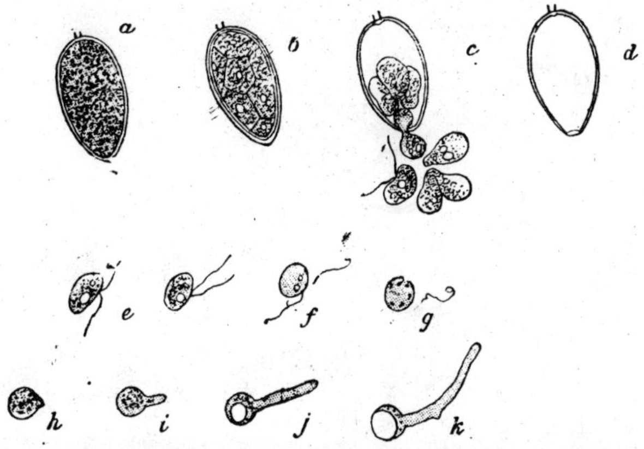

Fig. 17—The stages of germination of one of the conidia of *Phytophthora*. (*a*) the ripe conidium in water; (*b*) protoplasmic contents breaking up into blocks, which separate and escape (*c* and *d*) as minute kidney-shaped zoospores (*e*), each with two cilia; *f* and *g*, the zoospore coming to rest and losing its cilia; *h*, *i*, *j* and *k*, successive stages of germination of the zoospore. Highly magnified.

the tiny hypha developed from the zoospore dissolves its way into and through the cell-wall of an epidermis cell (Figs. 18 and 19), and finds its way into the living substance, quickly absorbs and feeds upon this, and starts a new mycelium, the branches of which spread with enormous rapidity between the cells of the leaf, and carry on the work of destruction already described. In certain cases the conidia germinate in a slightly different manner;

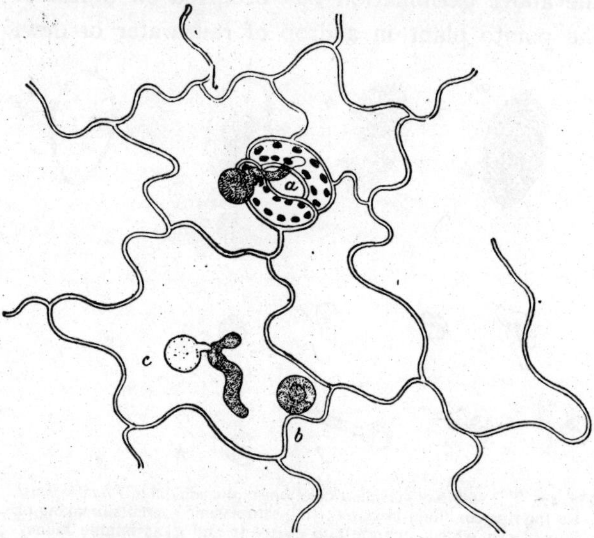

Fig. 18—Germination of zoospores of *Phytophthora* on epidermis of potato. At *a* the germ-tube is entering a stoma; at *c* it bores directly through the cell-wall. Very highly magnified.

but since the final result is the same, it is not necessary to describe the more abnormal process in detail here. A point of much greater importance is that the zoospores can also directly infect the young potato-tubers, the germ-tubes boring through the delicate skin of the latter, if they lie on the surface of the soil, or if the conidia are washed down by rain between the particles of soil. This infection of the tubers only succeeds, as a rule, if they are still young, since the coats of older potatoes are thick and corky, and resist the inroads of the hyphæ. It is not surprising to find that different varieties of potatoes show differences in this respect.

If one reflects that millions of the above zoospores may be developed in a few hours in moist, warm, summer weather, such as we often have in July, and such as is frequently accompanied by thunderstorms, the astonishingly rapid spread of the disease is easily accounted for. In an ordinary potato-field the abundant foliage, wet with rain and full of juices such as would favour the growth of the mycelium, is swayed by the wind, and leaf flaps upon leaf over the whole area; quite apart from wind-blown conidia, the active zoospores can soon spread from any one centre, and at once infect new leaves. In a few hours fresh disease-spots are developed, each putting forth new crops of conidia, which again germinate and send out zoospores, and so on. What wonder, then, that the farmer or gardener having failed to detect the first few diseased patches, or leaves, or even plants in his field, awakes some day to the conviction that many of the plants are infected; perhaps finding two or three days later that every plant in the field is attacked. It appears to be so sudden, that he is apt to ascribe the fatal attack to some general influence, and vaguely thinks of the clouds, the thunder, electricity, and so on; but

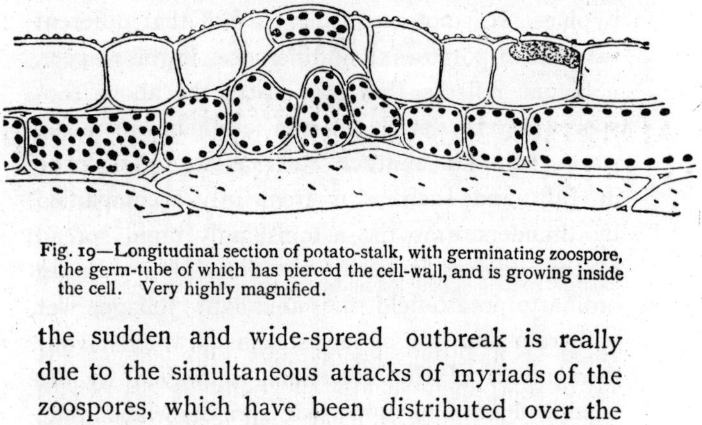

Fig. 19—Longitudinal section of potato-stalk, with germinating zoospore, the germ-tube of which has pierced the cell-wall, and is growing inside the cell. Very highly magnified.

the sudden and wide-spread outbreak is really due to the simultaneous attacks of myriads of the zoospores, which have been distributed over the field during the preceding days or weeks. The obvious connection between the pestilence and the weather is, that the latter favours the rapid development of the fungus.[1]

[1] The ingenious suggestion has been made that electrified spores may be rapidly precipitated during certain periods of atmospheric disturbance, much as other particles of solid matter are thrown down at such times.

Passing to the consideration of what means are available to lessen the virulence of this awful pest, which has caused misery to thousands of people in its time, and involved the expenditure and loss of hundreds of thousands of pounds, it is to be seen that no treatment is likely to succeed which is not based on the scientific knowledge of the habits and structure of the potato and the fungus. The case amounts to a rational application of botanical science, and without this, agriculturalists are likely to go on expending much fruitless time, labour, and money, as they have done in the past. It is obviously useless to look for a "cure" for the potato-disease, if by cure is meant some substance which can be supposed applied to the diseased crop, and stopping the progress of the disease; and yet it is astounding how many people seem to be still credulous on such matters.

If we take all the known facts about the potato-disease into consideration, it is clear that our only hope of combating successfully with the pest is to (1) take all precautions that tend to prevent the fungus from establishing itself in the potato-fields; and, (2) do the best possible to select such varieties of potatoes as offer most obstacles to the establishment of the fungus.

To put the extreme case, it evident that if we could start with clean soil, and plant tubers absolutely free from dormant mycelium, there is no reason to expect the disease, *unless spores are carried later into the field.* Now, how are spores carried? They are blown by the wind; they are flapped from leaf to leaf in wet weather; and they are carried on the clothes of men, the fur of animals, the feathers of birds, and even by insects. We cannot directly influence all these means of transport, but we can control some of the most dangerous of them—*e.g.* prevent people who have just passed through badly-diseased foliage passing through a field of healthy potatoes.

Much may be done by selecting the "seed" potatoes, which should never be saved from plants of a diseased crop. Of course only a madman would allow the diseased haulms of the potatoes to lie about on the ground, or to be placed on the store-heaps, or in any way to endanger the crop, if he understood the foregoing facts; it is because agriculturalists do not sufficiently appreciate the power of this invisible little enemy, that so much apparent recklessness is shown in this respect. The haulms should be removed from the ground at once, and placed on manure heaps that will ferment strongly; during the process of decomposition the spores and mycelium, which have been described, will be destroyed, at least to a large extent. Of course it is safer to burn all the diseased parts, but that is less practicable in many cases.

In spite of all precautions it may happen that a field of fine potatoes begins to show traces of the disease in patches, the fungus having been introduced in a stray tuber, or from a neighbouring garden or field, and that the whole crop is attacked before the presence of the malady is discovered. In such a case every effort is to be made to save the tubers, and one or two statements may be given as to the chief precautions directed to that end. In the first place the young tubers should be covered with several inches of soil if, as often occurs, they are close to the surface; this decreases the chances of the numerous conidia being carried to the tubers by the rain.

Numerous researches have shown that such substances as poisonous salts, sulphur, &c., cannot be used with safety, even if they could be depended upon; it is quite true that corrosive sublimate, arsenical salts, and other mineral poisons kill the fungus, but they also injure the potato plant, and are dangerous in other respects. Many have supposed that the removal of the foliage and stems from the still living plant is advisable. This needs a little thinking about. In cases where the disease appears late, it is well to remove all the foliage and stems, and little harm results; but if the young potatoes are only just "swelling," the removal of the foliage means that a short crop of small tubers only will be gathered, for it must never be forgotten that the starch, &c., which form the potatoes, has to come from the leaves. Much judgment is required to decide in given cases, and I cannot here go into all the *pros* and *cons* of the matter, as different complications arise in different cases, the clue to the unravelling of which is only to be sought in a proper knowledge of the physiology of the plant.

Finally, it should be pointed out that there is much reason for hoping that improved varieties of potatoes will yet be found to yield heavy crops in spite of the disease; something has already been done in this direction, but unfortunately, the raising and testing of new varieties is still often conducted in too haphazard and unscientific a manner; and until systematic experimental cultures are carried out on a proper scale, the development and realization of this hope must be slow. It is stated that the thick-skinned, redder varieties, growing on light soils, develop more starch and suffer less than do the white, thinner-skinned and less "mealy" potatoes, especially on heavy soils; and we have, moreover, some explanation of these facts which will be useful in directing further research.

CHAPTER VI.

THE "SMUT" OF CORN.

It is a very common experience to see some of the ears of wheat, barley, or oats, in a field of grain, presenting a blackened, shrivelled appearance, in sharp contrast to the golden gleam of the ripe, full ears around; and if one gathers such "smutted" specimens, they are found to be withered, and full of a soot-like powder (Fig. 20). This powder consists of myriads of minute, dark, spherical spores of a fungus called *Ustilago*,[1] a word having reference to the burnt or charred appearance of the diseased grain.

The particular species which produces the com-

[1] The particular species here considered is *Ustilago segetum*, Bull (*U. carbo*, Tul.); but it may be pointed out that most of the *Ustilagineæ* behave very similarly, causing much damage to all kinds of corn, maize, and other grasses, and many garden and other plants—*e.g.* violets, figs, anemones, scabious &c &c

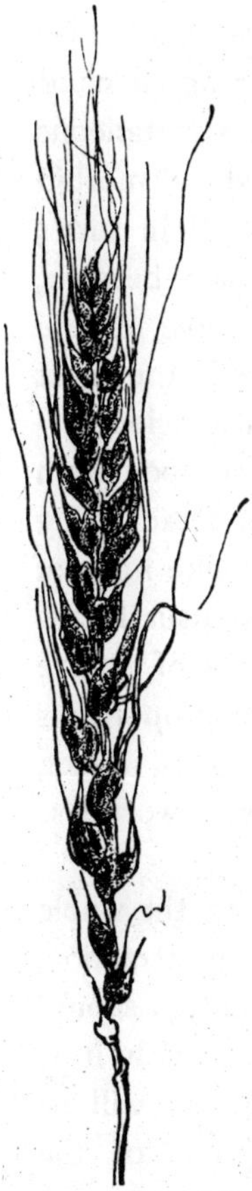

mon "smut" or "brand" in our corn-fields is not only very widely spread in Europe and elsewhere, but it is also found on a considerable number of grasses which grow wild, or cultivated in our fields, or on the sides of roads, &c.; a fact which has caused great trouble and confusion. Moreover, it belongs to what may be called a very "ancient family" of fungi, and one which has created a great stir in the world during historic times; for these "smuts" and "brands," and their allies the "bunts," or at least the diseases so named, were known to the Greeks and Romans, and have been the subject of many a learned disquisition on the marvellous, by those who used to write of the malevolence of Nature, the blighting influences of clouds and thunderstorms, visitations of anger on

Fig. 20—Ear of barley infested with *Ustilago segetum*, showing the grains bursting and disclosing the black spores of the fungus. *After Tulasne.*

the part of offended deities; and, nearer our own times, they have been over and over again cited as witnesses to the possibility of "spontaneous generation." They have been regarded as "morbid products," as animals, and as fungi, all in turn; and a classical interest attaches to them because, apart from their older history, they have of late years supplied the material for some of the most brilliant discoveries of modern biological science.

The reasons for the attention bestowed upon them are simple; every ear of corn in a spike attacked by "smut" is converted into a black soot-like mass of spores; and since each plant attacked usually has the "smut" developed in all its spikes, of course the loss may be enormous, when large proportions of the crop are assailed. In extreme cases, in fact, as much as half the crop may be rendered worthless by the disease.

Then the mystery which attached to the whole subject of these "smuts" was in itself an attraction, and another reason why so much attention should be directed to them. This mystery, which it is our purpose to show is now cleared up, will be best illustrated by a brief description of the farmer's experience of the disease.

The "seed" grain is sown in due season, germinates, and gives rise to the well-known green "blade," and all goes on well to all appearance, until the grain begins to fill and ripen. Suddenly, as the grain is approaching the time for harvest, the farmer finds a larger or smaller proportion of the "ears" taking on the sooty look so characteristic of "smut" (Fig. 20). Each bad grain is filled to bursting with minute, dark, powdery spores.

Ordinary microscopic observation yielded little more information, and for very many years even those who had satisfied themselves that the dark powder really does consist of spores, could find no traces of fungus-filaments. It was not till about 1860 that we really knew of the existence of a definite fungus-mycelium in the diseased corn-plant; but about that time, repeated and improved investigations had demonstrated the existence of a few scattered and extremely minute hyphæ, here and there, and it was also discovered that the dark spores, although free and powder-like when ripe, really spring from the ends of almost immeasurably minute hyphæ in the seed and other parts of the flowers and fruits.

Information was also coming to hand in other directions. It had hitherto been a mere matter of conjecture as to how the fungus got into the host-plant at all; analogy from the behaviour of other fungi led many to suppose that the black spores fell on to the leaves or flowers of the corn, and put forth fungus-filaments into the plant. Of course the only way to prove this was by experiments; but although many attempts were made, no one had so far been able to infect corn plants artificially and systematically, until at last, putting many things together, and trying various experiments, it was found that the fungus enters the seedling when it is still very young, and then grows up with the growing plant, and does not show itself until the fruit begins to ripen.

I have given this rapid preliminary sketch in order to lend clearness to what follows; but before going further, it will be necessary to say something about the spores.

It has long been known that the sooty grains are spores, which will germinate in water. If a little of the powder is placed under a high power of the microscope, each granule is seen to be a spherical or spheroidal cell, with an olive-brown coat, on which extremely minute points can be detected (Fig. 21). It would take from 150 to 200 of these spores, side by side, to measure one millimeter, and therefore about twenty-five times that number to measure one inch. Their extreme lightness when dry enables them to float about in the air, and when we reflect how tiny they are, and that countless millions of them may be formed on a small patch of corn plants, it is easy to explain the ubiquity of the fungus. These spores may be preserved dry for many years, and still

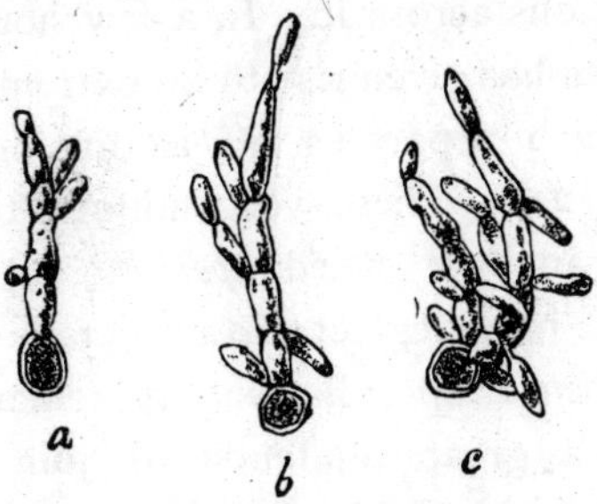

Fig. 21—Spores of the *Ustilago* germinating, and giving off sporidia from the germinal tubes or *promycelia*. (*a*) germinating in water only; *b* and *c*, in nutritive solutions. Very highly magnified. *After Brefeld.*

retain their power of germinating, and causing the infection of the corn and grasses.

If we place a few fresh spores in a drop of water, and keep them properly moistened and aërated, and not too cold or warm, they begin to germinate in from twelve to forty-eight hours or so, according to circumstances. And now comes the first puzzling fact in this eventful history. The outer dark covering of the spore bursts, as usual, on germination, and the protoplasmic contents, enveloped in a very delicate cellulose membrane, emerge as a short, colourless and nearly transparent tube (Fig. 21); this little tube grows until its length equals about five times the diameter of the spore, and then it becomes jointed by about three partitions across it. In a few hours, each of the segments has given rise to an extremely minute process, which separates off as an independent organ (Fig. 21). These very minute, translucent delicate organs are called *sporidia*, or spore-like bodies. We may neglect certain details and peculiarities of these sporidia, simply remarking that they show a great tendency to join in pairs; because the important point for us is, that after years of hard work on the part of investigators, it was discovered that these sporidia infect the young corn plants. The infection is not accomplished directly by the spores, then, but the latter produce the little tube from which the sporidia are budded off.

It was not until 1866, or thereabouts, that the proof of infection by means of these excessively minute sporidia was forthcoming, and not till much later did we learn all the details. What is now known may be shortly stated as follows.

If a *sporidium* comes in direct contact with a very young germinating seedling of wheat, barley, oats, &c., just about the time when the little root is emerging from the seed and entering the soil, and the still minute green "blade" has not yet made its appearance in the light and air, the infection takes place as follows.

The sporidium, in contact with the moist tissues, emits an extremely delicate germinal tube, the end of which bores into the young cells; this can only occur so long as the tissues are still quite young and tender—all danger of infection is over as soon as the cell-walls of the seedling are hard and thickened. Once inside the young seedling, this infective germinal tube bores its way from cell to cell until it reaches the growing apex in the bud of the shoot (Fig. 22). Here it takes up its abode, growing with the bud, and passing branches into each young side branch as it is formed, and thus gradually permeating the whole of the plant. The most remarkable part of the whole story is, that this fine parasitic mycelium can go on living for months in the tissues without our observing anything wrong, and for some years this was looked upon as a suspicious circumstance, and the experiments were only half credited, as it were.

But, simply as a matter of fact, we now know

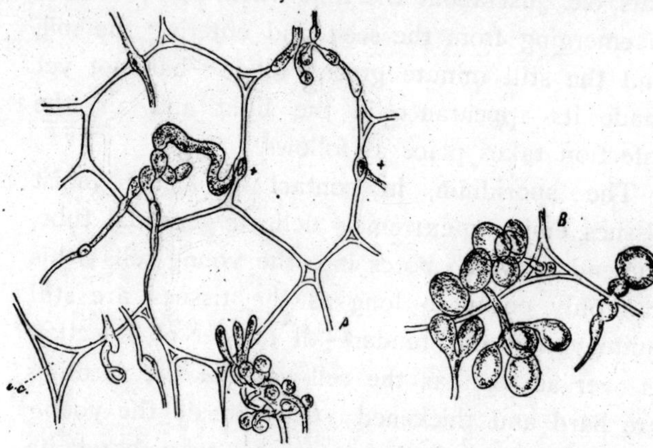

Fig. 22—Extremely thin section of the tissues of a grain of corn affected with smut, to show the intercellular mycelium of the *Ustilago;* at × the hyphæ are seen in the middle-lamella. At *B* the ends of the hyphæ are developing into spores. Very highly magnified. *Partly after Frank.*

many fungi which are able to go on living in the tissues of their hosts without forthwith destroying them—nay, such parasites may even stimulate the tissues to greater activity and vigour for the time being, and such is the case with the *Ustilago.*

The mycelium of the *Ustilago*, then, simply

keeps pace with the growth of the corn plant through the early summer, and by the middle of July it can be found in the various young organs, especially at their very tips. Further back, in the older and lower parts of the plant, it may be that none of the mycelium can be found, for the fungus hyphæ often die off behind and leave no traces. In each of the flower buds, however, there will be a copious supply of the mycelium, and now approaches the time when it is, so to speak, to be rewarded for its weeks and months of patient and subordinate growth. Up to this point the mycelium has been in abeyance, simply creeping onwards and upwards to establish itself; now, however, when all the energies of the corn plant are to be engaged in swelling out the young grain with starch and other valuable food-substances, the parasite becomes master of the position and revels in these stores, rapidly growing at their expense, and forming a dense meshwork of fine hyphæ inside the swelling young grain. Nay, its activity even stimulates the flow of food-materials from other parts of the corn plant, for it can use them up faster than the normal healthy grain could do, and consequently more supplies flow in, and the diseased grains swell up to nearly twice the size of the healthy ones.

The discovery of these phenomena cleared up much of the obscurity and mystery in which the disease had hitherto been involved. It became clear that the brown spores, ripening and scattering as the grain ripens, are harvested and threshed with the corn, and even garnered with the grain, so that the farmer sows both together in the spring or other sowing-period; then, as the seed germinates, so also do the spores attached to the seed-coats, and it would be strange if some of the sporidia produced from the spores did not meet with suitable points of attack as the tender seedling emerges into the damp soil. Moreover, these discoveries also explained why the outward and visible signs of the disease make their appearance so suddenly, giving rise so naturally to the metaphorical idea that the fields were blighted at one fell stroke by the "brand" or "smut"; and, again, it was at once seen why people had failed to infect the corn directly, as it stood. The whole mystery, in fact, began to break up and wear the aspect of a consistent and explicable story, as soon as it was understood that the mycelium of the fungus lay hidden in the tissues of the growing corn, awaiting the period of its triumphant conquest when the grain begins to draw to itself the stores of valuable substances which the plant had been so long preparing. It is beyond my purpose here to show how the demonstration of this long "latent period," on the part of the parasite, also threw a bright light on many analogous facts known in connection with other parasitic diseases.

But although so much had been accomplished, there was still more to follow, as will be shown shortly. One practical consequence of the above discoveries was soon apparent. It became evident at a very early stage in the investigation, that if the disease is started by the agency of germs (using this term in its general sense) sown with the corn, then it ought to be possible to remove these germs from the hard, tough, ripe grains of corn before they are sown. This kind of reasoning led to the establishment of the many modes of "dressing" grain which have been employed at various times, and which have now been so far perfected that one of these diseases—the "bunt" of wheat—has been nearly expelled from our corn-fields.

The methods employed are chiefly three. First,

it was argued that as the germs attack the young seedling, it ought to be possible to so coat the seed-grain with some poisonous chemical that the infecting hyphæ should be killed before they could enter the tissues; in the first instance, indeed, the experimenters hoped to kill the brown spores themselves, before germination had commenced; but further experience showed that the hard coats of these spores protect them too well for that. In an empirical sort of way, experiments proving the general fact that the disease is an *infectious* one were made many years ago, and even in 1820 it had been shown that grain gave fewer diseased ears if it was washed with copper sulphate before sowing. The usual methods are to soak the grain for a short time in a weak solution of copper-sulphate in water, and then to remove and rapidly dry it for sowing. It is argued that sufficient of the mineral poison adheres to kill the sporidia and germ tubes as they form, but not enough to injure the seedling wheat, &c. The copper sulphate is gradually decomposed by the lime in the soil,[1] but not before it has done its work. In some districts the

[1] The copper sulphate thus also acts chemically on the soil, making certain substances more available for absorption by the root-hairs.

grain is dressed with stale urine and lime, or with lime alone. Common salt, sodium sulphate, and even permanganate of potash and other disinfectants have also been employed for dressing.

A second method proposed was to simply and thoroughly wash off the spores, by rapidly agitating the grain in clean water: it was suggested that the spores, being light, would float to the top of the water and could be drawn off, while the soaking of the grain would do no harm, but might even hurry germination. Although it is quite possible to wash off most of the spores in small experiments, however, it does not seem to be so feasible where large quantities of grain are concerned, as on big farms.

The third process for ridding the grain of the disease also depends on the assumption that the spores can be mechanically removed; and it was proposed to subject the seed-grain, before sowing, to rapid and violent blasts of air, the argument being that the spores can be blown forcibly from their attachment to the coats of the grain. However successfully this process may have been carried out, it is found that many grains are liable to be injured unless great care is taken.

Of course, in any of the "dressing" processes

precautions are to be taken, as experience directs; the grains must be steeped in plenty of liquor, as otherwise the swelling of the deeper ones drives the upper ones above the level of the fluid; time is also required to enable the liquid to displace the film of air clinging to the coats and hairs of the grain. But the steeping must not be too prolonged, nor the dressing solution too strong, or germination is retarded—obviously a disadvantage in most cases, though I believe that it has been regarded as no drawback in some soils and seasons. Rapid singeing of the grain has been attempted, but the spores are not easily reached this way, and when dry they will withstand very high temperatures for a short time; freezing, as ordinarily experienced, has no deterrent effect.

As time went on, and experience was gathered, it was found that even the best and most successful "dressings" only diminished the proportion of diseased ears, and this more especially in the case of the "bunt." Even in fields of the best dressed corn, the "smut" was apt to make its appearance, though by no means to the extent formerly experienced; "dressing" certainly did much, but for some reason or other the "smut" still made a show in the struggle for existence.

Among other reasons for this, it appeared that the *Ustilago* in question exists as a parasite on certain wild and cultivated grasses other than our cereals, and that spores from these may reach the soil when the "dressing" is becoming exhausted. For instance, this particular fungus is found on the rye-grass and fescue of our meadows, and on a dozen or so of common way-side grasses in addition.

Nevertheless, there were many difficulties in the way of referring all the discrepancies between theory and practice to the invasion of spores from other grasses. Among other experiences, it was insisted upon by many that fresh manure was a fertile source of danger; in some cases this was explained by showing that "smutted" ears of corn occurred among the straw mixed with the manure, and experiments were made proving that the spores of the *Ustilago* can not only pass uninjured through the alimentary canal of a cow, but that they germinate all the better afterwards.

But in 1883 an entirely new gleam of light was thrown on all these difficulties by the publication of Brefeld's discoveries on this subject, and which showed once again that the scientific methods of Biology must be rigorously followed if we are to win the long fight with Nature to which we are irretrievably committed.

Brefeld showed that if the spores of the *Ustilago* are sown in watery extracts of manure, or other vegetable substances such as manure and soil contain, their mode of germination differs from that described in some important particulars—important because, as we shall see, it is just in such liquids that the spores are apt to germinate in corn-fields, unless the cultivation is so high and good that the farm is particularly clean.

In such fluid extracts of food-materials, then, the spores commence to germinate as in water, putting out the little germ-tube from which the sporidia begin to sprout much as before (Fig. 21); but instead of stopping at this stage, the sporidia begin to sprout forthwith on their own account, and each sporidium soon gives rise to many others (Fig. 23). Each sporidium soon begins to put out a little bulge from its surface, this little bulge grows till it is as large as the sporidium, and then it falls off, and *repeats the process of sprouting*. Moreover, as each series of these little sprouts is completely formed and separated off, every sprout sets about doing the same; and their progeny repeat the process, and so on almost indefinitely, so long as food-materials hold out. Brefeld calls each of the little sprouts a *conidium*, a word which refers to its minuteness—like a tiny particle of dust.[1] Now, it is easy to see that supposing we start with one of these conidia in a drop of liquid manure, and it

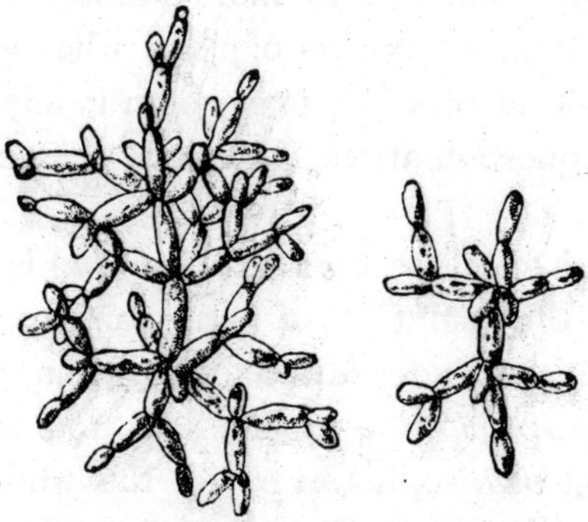

Fig. 23—Two colonies of sprouting conidia, developed by continued budding from one of the sporidia of Fig. 21. Each of these daughter conidia is capable of repeating the sprouting in a nutritive solution, or of directly infecting the young tissues of a corn-plant. Very highly magnified. *After Brefeld.*

puts out a new sprout which separates as a new conidium in half-an-hour, and at the end of a second half-hour each of these two conidia has produced another one, making four in all; at the end of a third half-hour each of the four has

[1] Each is about $\frac{1}{100}$th of a millimeter long—*i.e.* it would require about 2,500, end to end, to make a row one inch long.

formed another, making eight; and if the process goes on equably, the progeny doubling their numbers every half-hour—it is easy to see, I say, that the numbers produced soon amount to very many thousands, and that the products in twenty-four hours would amount to more millions than the mind could form any idea of; and when we reflect that this kind of thing may go on in any manure heap, the question arises, How does any corn-plant escape?

For each of these "germs" (*conidia*) is capable of giving the "smut" to a corn-plant, provided it meets with the tender surface of the young seedling *at the proper moment* as it comes out from the grain. In this last proviso, however, lies the whole secret —the only chance usually afforded to the parasite is that it shall have its sporidia or conidia ready on the spot and at the critical moment, to send their little infecting hyphæ into the tender tissues of the young seedling, at the part where the root joins the "blade" portion; unless the germs are there, and in contact with that particular small area at the right moment, there is no chance for them. It is no use their arriving at the place later, because experiments show that they cannot enter when the tissues are hardened, and although we know they have the power of penetrating into other of the embryonic tissues of the corn-plant if put there artificially, their chance of reaching these in nature is practically *nil*, because these tissues are protected in so many ways and so thoroughly in the bud.

It seems to me that we have, in these recent discoveries, a very fair explanation for the mystery which has hitherto hung around the story of "smut." It is only because the *Ustilago* produces such inconceivably large quantities of these minute spores and still more minute and numerous conidia, that it can dodge, as it were, all the chances against it. The waste of conidia must be prodigious so far as *numbers* are concerned, but this is compensated to some extent by the *minuteness* of each one. Botanists have long been aware of the enormous waste of pollen-grains which occurs in every pine-forest—millions are blown away for every one that finds its destination in the micropyle of an ovule; but the numbers here concerned are probably small compared with those above referred to, which only find their equal, perhaps, among the *Bacteria* and similar germs.

In all such cases, however, it must be remembered, firstly, that the only chance the organism has to establish and maintain its position in the struggle for existence with other organisms and the whole of inorganic nature, is to go on producing these apparently extravagant quantities of units of chance; and, secondly, that in the numerous and continuous transformations in nature, however wasteful the lavishness may appear to us, the materials are not only not lost, but are repeatedly being used up in various forms.

CHAPTER VII.

"BLADDER-PLUMS," OR "POCKET-PLUMS."

AMONG the numerous parasitic diseases which affect plum-trees in Europe and America and elsewhere, none is more remarkable than the one chosen as the subject of this chapter. Its chief symptom is the curiously malformed developments which replace the normal fruit, and which are caused by the fungus *Exoascus*,[1] which attacks the very young fruits just as they are beginning to swell, and causes their tissues to change their characters almost entirely; the consequence is, instead of a ripe luscious plum, with a properly hardened "stone" containing its seed, we find an unequally formed, dirty greenish or yellow, hollow and tough body, called a "pocket" (Fig. 24).

[1] The particular species is *Exoascus Pruni*. The word *Exoascus* refers to the peculiarity that the *asci*, or bags containing the spores, are exposed on the surface of the plum.

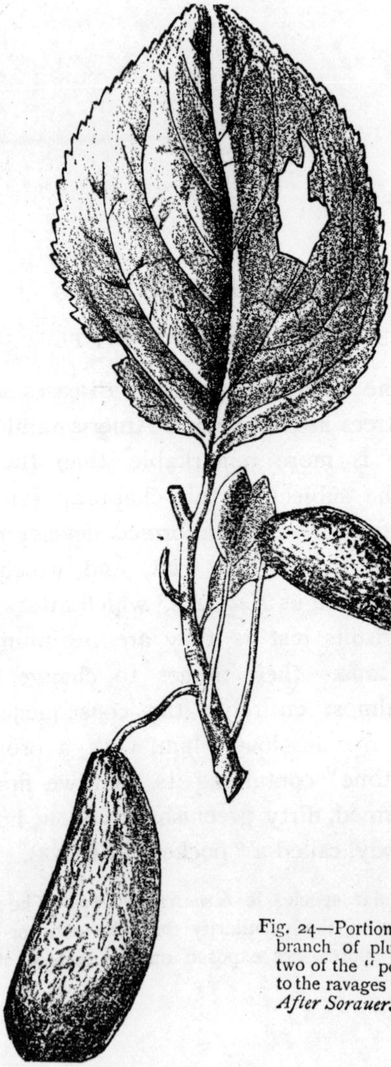

Fig. 24—Portion of a fruiting branch of plum, showing two of the "pockets," due to the ravages of *Exoascus*. *After Sorauer*.

The disease has been well known from of old, and much difference of opinions is apparent among the earlier writers as to its nature and cause, some confounding the "pockets" simply with stoneless varieties of the plum, others referring them to abortions or hypertrophies due to cold or wet, and others again regarding them as galls produced in consequence of the bites of insects. More recent investigations, under improved conditions, have placed the malady beyond doubt in the category of fungus diseases, and it is satisfactorily established by the discoveries of De Bary and others that the action of the mycelium of the *Exoascus* is the exciting cause of transformation of the tissues.

If we remove one of the young "pockets" in May, when they usually begin to develop, we shall find it a somewhat elongated, curved, fairly pliant mass, of a yellowish colour, and more or less wrinkled or grooved (Fig. 24). Specimens taken at a later date may be redder, and will probably be covered with a grayish "bloom" or powder; later still they will be spotted and rotten with mould.

Cutting across the "pocket" shows that it is hollow, and it is doubtless partly due to the fact that insects may make their way into the rotten mass, that the idea arose that these structures were of the nature of galls; the section also shows that no "stone" is developed, and the shrivelled remnant of what would normally have become the seed may be detected in the cavity. The walls of the "pocket" are formed of what should have ripened into the soft, succulent "flesh" of the plum; but they are now tough and useless.

In this last-mentioned tissue, just beneath the shrivelled skin, the microscope reveals the presence of exceedingly delicate fungus hyphæ, jointed, and branching in all directions, and running especially along the course of the fibrous strands, which pass up from the stalk into the plum. After filling up the tissues of the young "pocket," and growing with it, the ends of many of the hyphæ begin to turn outwards, between the cells forming the epidermis, or skin; and soon afterwards the tips of these hyphæ break through to the exterior, and stand off from the skin like so many nine-pins, only we must suppose the nine-pins in thousands and closely packed like the threads of the pile of a piece of velvet (Fig. 25). Moreover, the rough analogy to nine-pins of course only applies to the outward form of these projecting ends of hyphæ, each of which is a soft, colourless body composed of protoplasm enveloped by a delicate cell-wall. It is important to notice that the presence and activity of the hyphæ of the fungus really cause the malformation or "pocket," and this comes about somewhat as follows. The fungus hyphæ take food-substances from the living cells of the young swelling fruit, and this increases the flow of nutrient matters to each of the centres of attraction so constituted at each point where hyphæ are attacking cells. Since these hyphæ are in the parts which would have normally developed into the "flesh" of the plum, moreover, they divert the whole flow of food-substances into those regions, leaving what should become the stone and seed deprived of their proper supplies—hence they starve, shrivel, and die. But, it may be asked, how is it that the fleshy parts of the plum become thus altered if they receive all the food-supplies? The answer is, that of these food-supplies the greedy fungus absorbs, and partly consumes by respiration, just those sweet and pleasant substances which would have made the normal plum so delicious, leaving for the plum-cells the parts which become converted into tough, hard cell-walls, and insipid useless matters of other kinds.

Moreover, it is because the cells are no longer filled with the juicy and sugary materials of the normal plum, that they cannot fill themselves with moisture; hence the ugly, shrivelled, tasteless "pocket."

Returning to the fungus, the exposed club-shaped (or "nine-pin shaped") ends of the branches given off from the short hyphal cells beneath the outer skin of the pocket are called *Asci*, each *Ascus* being a bag[1] of thin cellulose, containing protoplasm and watery sap (Fig. 25). Very soon after its formation the *ascus* begins to undergo changes, in that its protoplasmic contents slowly divide up into about eight rounded masses, each of which surrounds itself with a membrane, and thus becomes a spore, termed an *ascospore* because it is developed in an *ascus* (Fig. 25).

It will easily be understood that the crowded *asci*, projecting from the surface of the deformed plum, give to the normally smooth skin an appearance of roughness or false "bloom," which is characteristic of the pocket at this stage; it looks, in fact, as if covered with a sort of short-piled velvet of a grayish or dirty yellow colour. As water collects on the surface, the *asci* burst in

[1] The word *Ascus* is practically the modern form of the Greek word for a bag.

thousands, and hundreds of thousands, and scatter the tiny *ascospores* to the winds and rains.

The germination of these spores is somewhat curious, and unlike anything we have yet met with. After lying for an hour or so in water, or

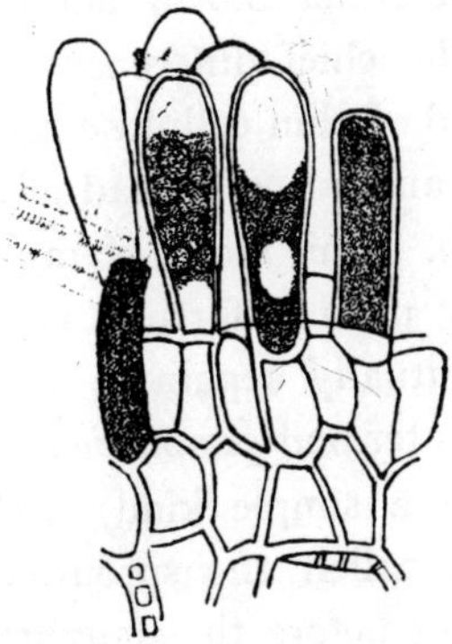

Fig. 25—Portion of a section, at right angles to the surface, of a "pocket-plum." There are hyphæ making their way between the cells to the exterior, where their ends swell up as asci, one of which already contains eight young spores. Very highly magnified. *After De Bary.*

dilute sugar-solution, or other suitable fluid, each spore is found to have swollen, and its broadly ovoid shape begins to change as follows: a minute papilla or bulge appears at one point on the periphery, and in the course of two or three hours this protuberance has become larger and larger, until it is as big as the spore. By this time the protuberance looks like a blown-out bladder attached by a narrow neck to the previous bladder —*i. e.* the *ascospore;* and in fact its development is in some respects not unlike the formation of a second minute bladder blown out on the side of an initial one, the chief difference being that the bladder is formed of thin cellulose membrane lined with protoplasm and water, instead of being inflated with air (Fig. 26). This mode of multiplication of a single cell, by putting forth a protuberance of itself which eventually separates as a second independent cell, is termed *gemmation* (a word meaning sprouting of a simple kind), and the second bladder may be called the sprout-cell. It often happens that long before the daughter sprout-cell is ready to separate from the parent cell, by the constriction and obliteration of the narrow connecting neck, the same process of sprouting is repeated at several points on both cells, and thus we have colonies of these sprouting cells and their progeny (Fig. 26).

It is a remarkable fact that the various species of plants which constitute what are called *Yeasts*, such as those of beer, wine, &c., consist entirely of cells almost exactly like these sprouting spores of *Exoascus*, and go on multiplying in this way through countless generations, producing complex changes in the liquids (wort, must, &c.) in which they are cultivated, and thus in fact converting these liquids into beer, wine, &c. The spores of the fungus we are considering will not bring about these fermentations, however.

In addition to reproduction by means of these

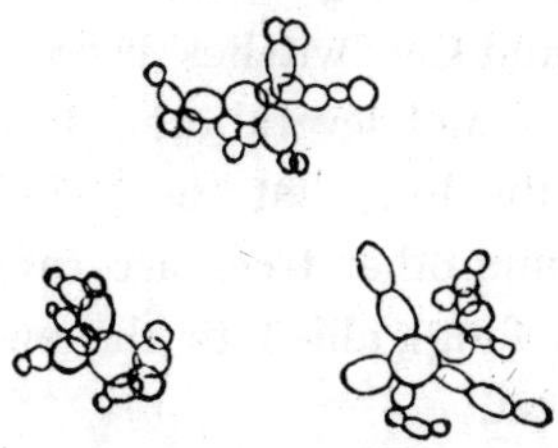

Fig. 26—Three groups of sprouting ascospores of *Exoascus*. Their behaviour is very like that of yeast-cells, and reminds one also of the budding of the conidia of the very different fungus *Ustilago*. Very highly magnified. *After De Bary*.

spores, which by germinating on the plum-trees convey the disease from one plant to another, the fungus is able to carry on its existence from year to year by means of the mycelium in the branches, and it is this circumstance which makes it so difficult to combat the disease. Obviously it is not sufficient to merely collect and burn the "pockets," and probably the only safe procedure is to prune back to the old wood. However, more will have to be discovered as to the exact behaviour of the *ascospores* and their mode of infecting healthy trees, before we can be sure of some of the ground we are now treading upon, and I know of few better instances than this of a disease which is interesting to everybody, and the amelioration of which concerns so many people, especially as allied species of fungi cause the "curling" of peach-leaves and the "witches' brooms" of cherries, birches, poplars, and several other trees. It should be noted, by the bye, that the "witches' brooms" of firs and some other trees are caused by a different class of fungi, allied to the one causing the "rust" of wheat.

CHAPTER VIII.

THE LILY DISEASE.

ONE of the most annoying pests that the horticulturist has had to trouble him of late years occurs in the form of a destructive "rot" on lilies, due to the ravages of a fungus ordinarily known as *Botrytis*;[1] and as this is a type of parasite which causes similar maladies in many other plants besides lilies, I propose to give a short account of it and its effects, including the marvellous transformations which it undergoes before completing its life-history.

In 1881 it was found that certain valuable species of lily, and especially the beautiful *L. auratum*, were being destroyed by a "rot" which attacked their buds and leaves (Fig. 27). At first small

[1] So named on account of the spores being in clusters resembling bunches of grapes; it is also called *Polyactis*, from the many radiating branches.

rusty-looking spots were seen, and these spread, the

Fig. 27.
Buds of lily attacked by the *Botrytis* of the lily disease. Nat. size.

flowers becoming distorted and spoiled. Quite

recently the disease has been noticed in many districts, and the following facts have been made out about it. The at first small, reddish-brown spots on the buds and leaves soon become larger, and if the weather remains damp they may spread over the

Fig. 28—Section through part of one of the badly diseased buds of Fig. 27, showing the tissues completely destroyed by the fungus, and the tufts of *Botrytis* issuing into the air. Magnified.

whole plant; the tissues of the leaf or bud collapse and rot away at these places, and a peculiar gray, mould-like fungus is found with its mycelium running in the dying tissues, and sending beautifully-branched hyphæ off from the surface into the moist air (Fig. 28). These hyphæ bear at the ends of their branches pear-shaped swellings, on each of which a dozen or more small egg-shaped conidia are produced, each conidium being attached to the swollen end of the branch by a minute peg-like stalk (Fig. 29). While still young these bodies are all white and glistening, giving the mould-like appearance referred to, but as they ripen the colour deepens to a pale brown hue. The dead and dying buds and leaves are covered with these structures in all stages of development, giving a peculiarly velvety appearance to the rotting buds, &c.

Botanists have long been acquainted with this form of fungus (*Botrytis*) on beans, clover, artichokes, vegetable-marrows, snowdrops, roses, and other plants, and have given various specific names to the slightly differing kinds met with; but since in most cases these fungi were found springing from the dead or dying tissues, it was assumed that something else had killed the plants, and that the *Botrytis* then grew on the decomposed substance of the tissues, much as *Mucor* lives saprophytically on dead vegetable remains.

It turns out, however, that these fungi are capable of leading a parasitic existence in addition to their habit of passing much of their lives as saprophytes, and I have chosen the particular form which produces the lily disease in order to explain

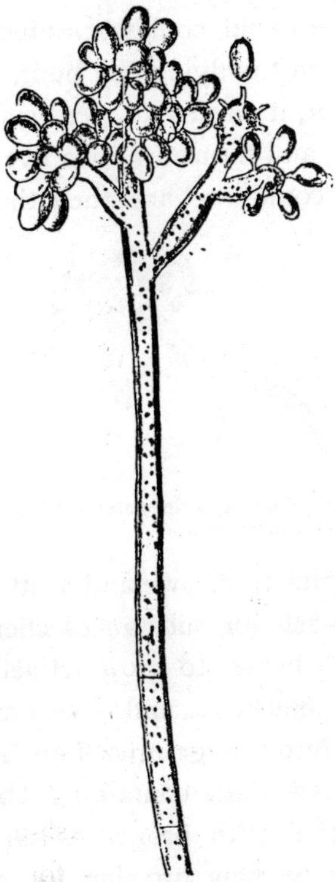

Fig. 29—One of the *Botrytis* heads removed and highly magnified, showing the conidia produced on peg-like projections on the branches.

how this curious phenomenon comes about, and

because it gives us some clue to the possible origin of parasitism generally.

If one of the ovoid conidia, produced in countless thousands on the dying lily buds, is placed in a drop of water, it begins to germinate in a few hours (Fig. 30), and forms a hypha five or six times as long as the conidium,[1] and then the whole dies.

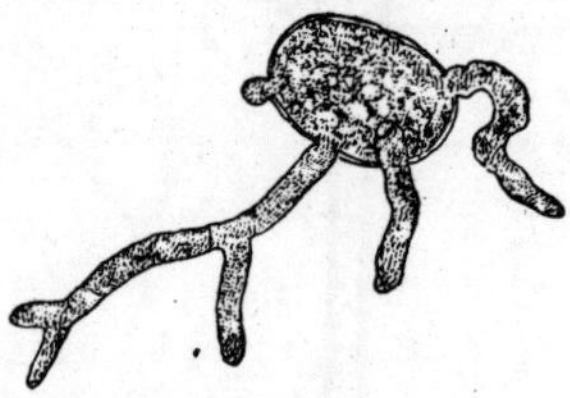

Fig. 30—A conidium germinating in water containing a little food material. Very highly magnified.

If, before it begins to die, we add a little sugar and traces of cigar-ash (or substances chemically like this), the hypha begins to grow actively, branches, forms septa, vacuoles, &c., and in two or three days has developed into a large mycelium (Fig. 31). If we place it in a flask containing abundance of the sugar-solution with proper ash-materials, this mycelium may be kept growing for months, and it is possible to obtain many ounces or even

[1] The conidium measures about $\frac{1}{50}$th of a millimeter long by $\frac{1}{70}$th broad.

pounds of this fungus-mycelium by taking proper precautions.

This mycelium sends many of its branches down into the liquid to absorb food-substances, and some of these branches are peculiarly close set, forming queer, tassel-like tufts, which stick to the glass or other solid objects; others of its branches join together and make the network of hyphæ more dense and felt-like; others, again, rise up from the surface of the mycelium into the moist air, and put forth short branches which swell at their ends and develop the ovoid conidia (Fig. 31) exactly like the one we commenced with.

There are many remarkable features of great scientific interest in connection with these and other structures, but as we cannot here enter into details which concern the specialist rather than the general reader, and which would need lengthy descriptions to make them clear, I must pass over most of them, and refer only to a phenomenon of considerable importance which has lately been discovered in connection with this fungus mycelium. Under certain conditions, if the fungus is well-fed and vigorous, the tips of the hyphæ excrete a substance which has the power of softening and swelling cellulose, the material of which the walls of the tissues of the lily as well as other plants are composed.

If, now, one of the conidia is sown in a drop of

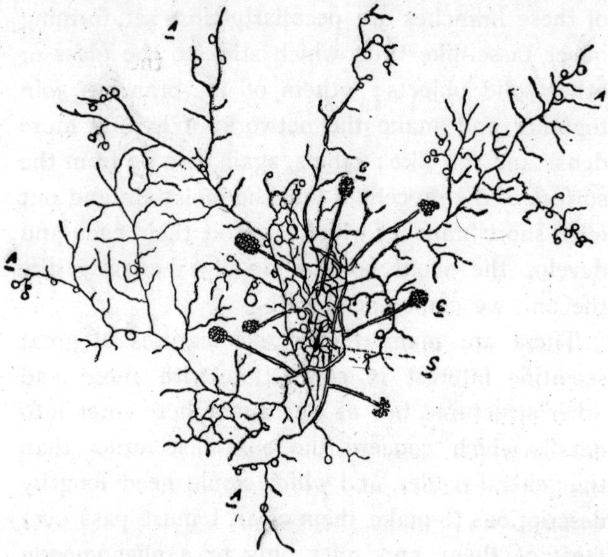

Fig. 31—A mycelium produced from one conidium, sown in a drop of nutritive solution. In addition to numerous branches (A) in all directions, it is already developing fresh Botrytis-heads C. Magnified.

water, as described above, and if a thin slice of lily-plant is placed in the drop, it is possible to see the hyphæ produced on germination pierce the cell-walls of the lily tissues and run through the whole slice, feeding upon the substance as they do so.

On the other hand, if a conidium is placed in a drop of water on a young living leaf of the white lily, or on a young flower-bud, the hypha put forth on germination soon begins to bore its way into the tissues (Fig. 32), evidently by excreting the above-named soluble substance, which enables its

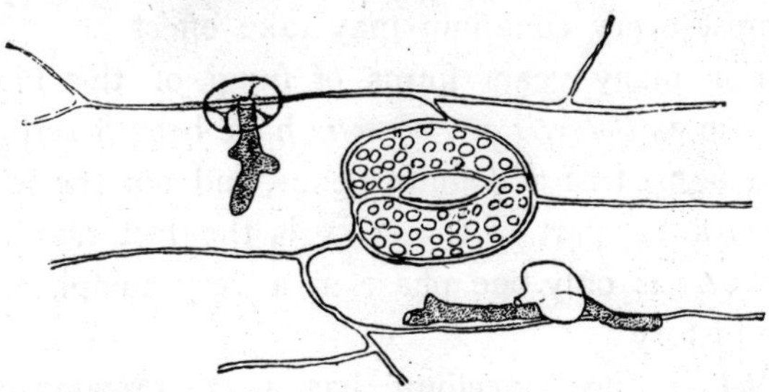

Fig. 32—Germination of conidia on epidermis of a lily. The germ-tubes at once bore into the cell-walls.

tip to pierce the cellulose cell-walls. Once inside, the hypha soon increases in length and thickness, and branches in all directions. At first the branches run in the cell-walls only (Fig. 33), but when the tissues begin to break down owing to the destruction of the walls, the parasite sends branches in all directions, and rapidly completes the work of devastation, and in a few days the lily-bud is a mere rotting mass of dead cells, in which the mycelium is luxuriating, and from which thousands of aërial hyphæ spring to the exterior and produce new conidia at the ends of their branches (Fig 28).

Immense destruction may be caused by this disease, since the conidia of the fungus are easily transferred by the wind, or by insects, &c., from plant to plant, and if the season is damp and the young growing organs correspondingly tender, almost every conidium may take effect.

For many years forms of fungi of this kind, known as *Botrytis* or *Polyactis*, have been described and figured as a definite genus, and not the least remarkable part of the story is the fact that the *Botrytis* is only one phase in a very complicated life-history.

When the mycelium has been growing for several weeks or months on nutritive substances in a flask, numbers of rounded or ovoid black, hard bodies are observed embedded in the felt-like mass of fungus. These *Sclerotia*, as they are called (from a Greek word meaning hard), are very curious structures. They are, in effect, balls of hyphæ produced by the repeated branching and knotting up of the ends of branches of mycelium. When a closely wound and twisted mass of small branches about the size of a large pin-head has been formed, the outer layers of hyphæ turn brown or black, and very hard, and the centre remains white and softer. The ball then constitutes a *Sclerotium*—*i.e.* a hard body. In the case of the lily fungus the sclerotia, which have only lately been discovered, are somewhat irregular in size and shape, and shining black, so

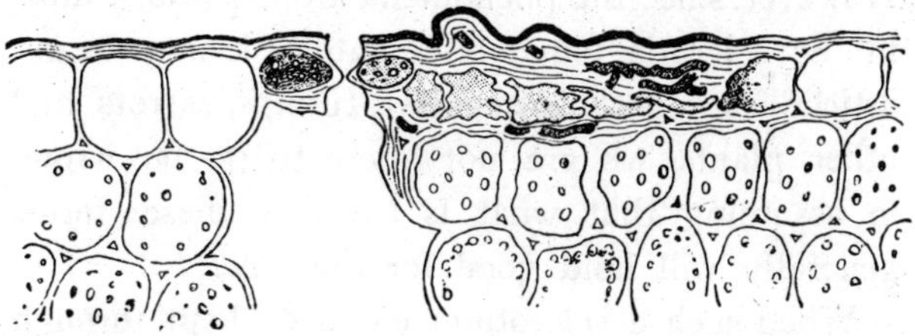

Fig. 33—Section through a lily-bud infested with the mycelium of the *Botrytis*, the hyphæ of which are seen in the swollen substance of the cell-walls. Highly magnified.

that they look somewhat like the droppings of mice, and might easily be passed over as foreign structures having nothing to do with the fungus.

But these sclerotia, of which we now know several different kinds, are not only produced by the fungus in question, but they are of the utmost importance to it; for when all the rest of the mycelium has been killed off by dry, cold weather, they remain uninjured through the frosts and snows of the winter, and the hard, black, outer coat of the sclerotium keeps alive the inner portion of knotted-up mycelium, until the advent of warm weather enables it to grow out again.

What follows has been established over and over again with respect to several of these sclerotia, although it has not yet been thoroughly worked out in detail for the fungus of the lily disease. However, since the phenomena are essentially alike in the sclerotia found on potatoes, kidney beans, artichokes, vegetable-marrows, turnips, carrots, and other plants, we are not likely to go far wrong in assuming that what is true for these forms generally will hold good for this one.

When such a sclerotium has been kept through the winter, and is put into soil in a flower-pot in the spring, and kept well watered, it may be found to sprout after a few days, putting out large numbers of the aërial hyphæ bearing conidia, and sending forth branches of mycelium in all directions. After some time, in the warmer season, one or more minute humps appear on the sclerotium, and gradually grow up as slender stalks above the level of the soil, and then spread out at the apex into a more or less funnel-shaped or trumpet-like structure which is called the *disc* (Fig. 34). This stalked disc-fungus, springing from the sclerotium, has long been known to botanists under the name of *Peziza*, a queer word referring to the fact that the fungus is based upon or sits upon something, many of these *Pezizas* springing directly from pieces of dead wood, leaves, &c. The

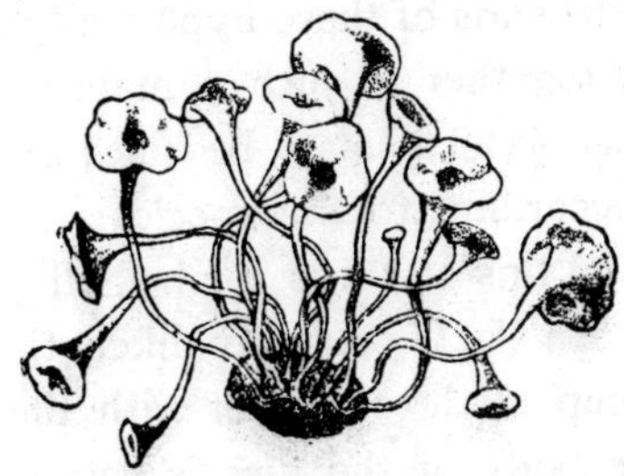

Fig. 34—A Sclerotium germinating and giving rise to trumpet-shaped *Peziza* discs. *After Brefeld.*

Peziza we refer to is remarkable in having a stalk and springing from the sclerotium (Figs. 34, 35), and in classifying the very numerous forms, attention has to be paid to these and other characters. The *Peziza* when quite mature is a very complex structure, and I shall have to give a short description of its leading features before proceeding further.

It consists, as we have seen, of a *disc* and a *stalk*, the latter springing from the sclerotium below the level of the soil (Fig. 35). The stalk simply consists of numerous parallel hyphæ of the fungus, and needs no minute description. The disc also consists of interwoven hyphæ, continuous with those of the stalk, and at first not unlike them; at maturity, however, the ends of these hyphæ all turn upwards, and grow together in such a manner that the top of the disc (which may be flat, or concave, or convex, according to the species) is composed of the rounded tips of these hyphæ, all pressed close together. In fact you may liken the disc to a plate or cup made of velvet with the pile uppermost; the body of the disc is composed of what would correspond to the felt-like basis or substratum of the velvet, and the upper surface or lining of what would answer to the pile, the upright threads of which would represent these upright hyphæ. Of course the analogy is a rough one only. The upright hyphæ are of two kinds: first, we have large numbers of barren filaments of the nature of ordinary mycelial branches, and secondly, evenly distributed between these, we find branches which are stouter and more club-shaped, and which contain in their cavities eight small egg-shaped spores (cf. Fig. 25). These tubular bags of spores are called *asci*—each one being an *ascus* filled with eight *ascospores*—and remind us of the *asci* of *Exoascus*.

Now it will readily be understood that the surface of the top of the disc, if it is flat, or the lining of the cup, if it is concave, is composed of the rounded upper ends of these erect and crowded parallel *asci* and barren filaments—the pile of the velvet. By a very pretty mechanism, depending on the absorption, &c. of water and the elasticity of the structures, the *asci* explode as their spores ripen, and so shoot the little *ascospores* into the air; and as many thousands of *asci* may be formed on one disc, some rough idea may be formed of the quantities of spores shot off day after day from the *Peziza*.

If a glass slip with a drop of water or nutritive solution is held over one of these bombarding discs, it is very easy to obtain the spores quite pure, and to watch their germination. They swell and emit a hypha as before, and this hypha soon branches and grows into a mycelium, and in a few days this mycelium may give rise to erect filaments, which branch, swell at the ends, and develop ovoid conidia, exactly like those we started with

—in fact the *ascospore* gives rise on germination to a mycelium which bears the conidia of *Botrytis*, whence we see that the so-called *Botrytis* is only

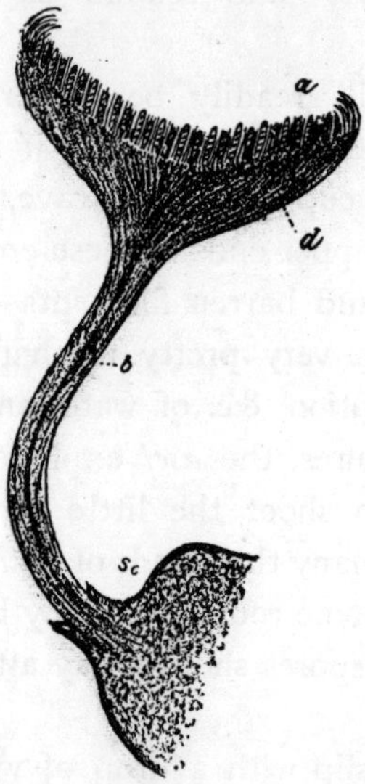

Fig. 35—Section passing through a sclerotium *sc*, and the *Peziza* to which it has given rise. *b* the stalk, *d* the disc, and *a* the asci of the *Peziza*. There are numerous paraphyses between the latter. Magnified, and slightly diagrammatic.

a phase in the life-history of a *Peziza*. I put it this way, taking the *Peziza* as the genus, because it is the more complex and highly-developed of the two phases.

It is not difficult to see what the course of the life-history of such a fungus will be in nature, now that we have elucidated the various phases as studied in the laboratory; and that these phases are passed through in the tissues of many plants is quite certain—it is also very probable in a large number of other instances.

Suppose a sclerotium in the neighbourhood of the plants in question: it germinates in the summer, on the ground, and produces the *Pezizas* as described above. These *Pezizas* go on day after day firing off the *ascospores* from their *asci*, and the little spores are carried by the wind and settle down, let us say on fairly rich soil near our plants. Here they germinate, and their hyphæ feed upon the nutritious organic matters, &c. extracted from the soil by rain, dew, &c. In a short time the mycelium thus produced may have crept along the soil far enough to reach one of the plants referred to, and the tips of the hyphæ would then bore into its tissues and so infect it; or the above-mentioned mycelium may produce the *Botrytis* or *Polyactis* form, the conidia of which directly infect the plant. In either case the result would be the same: the fungus, once inside the young and damp tissues, would play havoc, and at length send forth its hundreds of thousands of aërial conidia-bearing hyphæ (the *Botrytis* form), and shed millions of these spores on to the plants around.

Woronin has lately described one of these fungi, the conidia of which are scented, and attract insects, which then convey these spores to the stigmas of flowers. Here they germinate with the pollen-grains, and their hyphæ race the pollen-tubes down the style, and then develop the sclerotia in the ovary.

If one plant in a large bed was thus attacked, we see at once how the disease might rapidly spread over a considerable area. But it is highly probable, as demonstrated by recent researches, that this rapid and destructive spread of the parasite is only possible under certain conditions. In a dry summer the lilies show a few small spots only, but no very general or severe damage is done; in dull, cold, damp seasons, however, the flower-buds may be all destroyed, because the tissues remain for a long time in a soft, tender, "susceptible" condition, whereas the low temperature and diminished light do not check the fungus, or at any rate not to an extent more than is counterbalanced by the favourable conditions afforded by the damp and its consequences.

CHAPTER IX.

"ERGOT" OF RYE AND OTHER CEREALS.

THIS is a disease due to the ravages of a fungus on the fruits of sedges, grasses, and cereals, and has been especially examined in the case of rye, and which was for a long time in days gone by confounded with others of the numerous maladies which these widely-cultivated plants suffer from. It may be remarked, by the way, that there is little room for surprise that the cereals should be so especially noticed in these connections, for the reader will now understand that just such plants as these—crowded, herbaceous, and growing year after year in the same districts—offer peculiar facilities for the spread of an epidemic such as a parasitic disease. Moreover, the attention of man is of course soon drawn to anything abnormal in his corn-crops, and here we have another reason for knowing of so many diseases of the cereals.

The name ergot is given to a disease caused by the fungus now called *Claviceps*,[1] which apparently transforms the young fruits of various grasses, including barley, rye, wheat, &c., into curious elongated, purple-black, hard bodies of the nature of *sclerotia* (Fig. 36), which project from the ears. The name ergot is from the French, and has reference to the spur-like shape of the sclerotia. I say apparently transforms the young fruit, the truth being that the fungus lifts the dead ovary, or part of it, up as it grows, but does not occupy

[1] Named from the club-headed receptacles which bear the fructifications.

Fig. 36—An ear of rye attacked by Ergot (*Claviceps purpurea*), showing two large and several small sclerotia projecting from the grains.- *After Tulasne*.

its cavity in the same way as does the smut of corn, for instance.

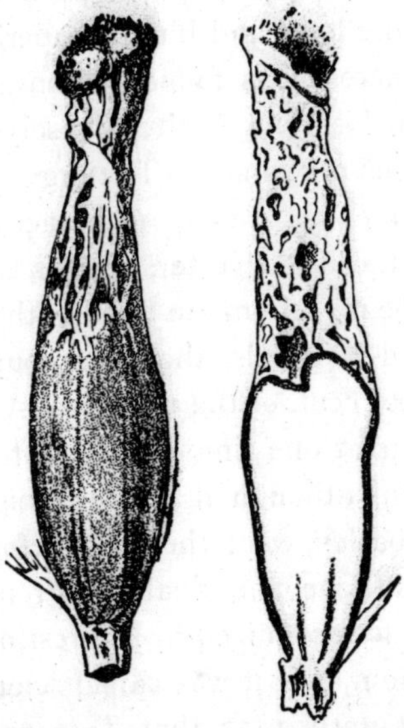

Fig. 37—Two grains of rye attacked by *Claviceps*, removed from the ear and slightly magnified. The lower half, resting on the pedicel, is the developing sclerotium or Ergot proper. This carries at its apex the now dried-up mass of fungus-hyphæ (*Sphacelia*) from which the conidia (Fig. 41—*a*) have been developed. In the right-hand figure the sclerotium is in vertical section. *After Tulasne.*

Since the ergot phase is the one best known, and most easily studied, I shall commence with it. If one of the black bodies projecting from the diseased ear (Figs. 36, 37) is removed just as it is ready to fall, it will be found to be a hard, curved, sub-cylindrical, and perhaps grooved body, about an inch or more long, and bluntly tapering at either end; if cut across it is found to consist of a dark firm coat, enclosing a white, densely compacted mass of fungus mycelium. The ergot is, in fact, a sclerotium somewhat similar in character to the one described in the last chapter, and, as in that case, the cells of the mycelium are filled with protoplasm and oily food-materials, the oil amounting to as much as 35 per cent. of the whole substance. This sclerotium was at one time supposed to be simply the bad grain, although it is much bigger than a grain of rye, barley, &c.; then it was found to consist of fungus tissue, and finally a French botanist named Leveillé recognized in it a resting condition of the mycelium, and it was called sclerotium.

It was not until 1852 that Tulasne discovered what became of these bodies after leaving the host-plant. On sowing a large number of the sclerotia, it resulted that, after at least three months or more of rest, a number of little hillocks gradually make their appearance all over the surface of the hard mass, and grow out in the form of little drum-sticks, so to speak, with plump round violet-coloured heads and rather thick white handles (Fig. 38). When this was first discovered the drum-sticks were thought to be parasites, but it was subsequently proved that they are composed of hyphæ which have grown out from the cells composing the sclerotium, and really belong to it: they are developed from

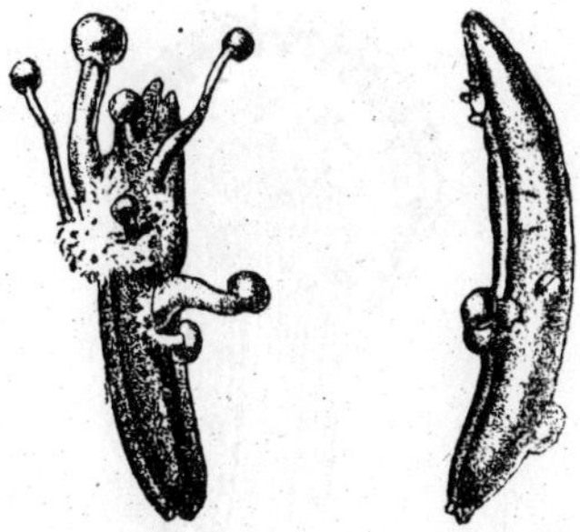

Fig. 38—Two germinating Ergot sclerotia. The one to the right has six stromata just bursting through; the one to the left shows seven stromata further developed. Natural size. *After Tulasne*.

the paler internal portions, and burst their way out through the thin hard dark-coloured coat.

In the course of time the pale violet head of the drum-stick[1] is found to be studded with numerous minute wart-like papillæ, each of which has a very minute hole at its apex (Fig. 39).

[1] Technically called a *stroma*—the bed or cushion in which the perithecia are immersed.

Sections through the head show that each of these holes leads into a somewhat egg-shaped cavity, placed vertically below the papilla (Fig. 39). We may roughly compare each papilla to a conical volcano, the small crater of which leads to a cavity below. In these egg-shaped cavities, which are

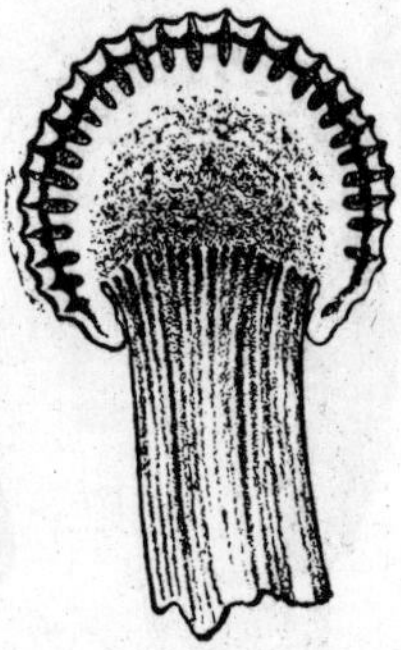

Fig. 39—Vertical median section of the upper part of a stroma, slightly magnified. Showing the pear-shaped perithecia sunk in the head, and each opening by a small orifice at the apex. *After Tulasne.*

called *perithecia*,[1] are produced large numbers of long, thin, tubular *asci*, not unlike those we found lining the cup of the *Peziza*, and, as there, a number of young *asci* in all stages of development are densely packed between these more mature

[1] The word *perithecium* may be translated a capsule or case surrounding and enclosing the *asci*.

ones; these *asci* spring from the base of the ovoid *perithecium*, and the long axis of all these structures is vertical to the surface of the round head of the drum-stick, so that the *asci* point towards the hole, or crater of the volcano.

In each of these *asci* are to be found from six to eight long, slender, colourless threads, which are the *ascospores*.

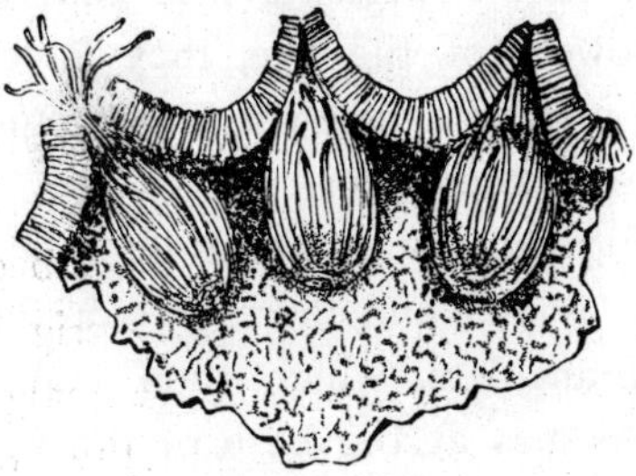

Fig. 40—Three of the perithecia more highly magnified, each full of long asci. *After Tulasne.*

On the ground, then, the black ergot sclerotium, after passing through a period of rest, gives origin to a dozen or more of the drum-stick-shaped *stromata* with violet heads and white handles, and in each head a large number of *perithecia* full of *asci* with their *ascospores* are developed. As the thread-like spores ripen, the bursting *asci* shed them through the small holes at the top of the papillæ, and they are carried about by wind, rain, insects, &c. In the natural course of events, it seems that phenomena have usually progressed thus far in June or July, that is just about the time when the grasses and cereals are commencing to flower.

It has been proved by direct experiment, that if these spores are brought into contact with the young flowers of the rye, they infect it in the following way, and give rise to a fungus mycelium in the base of the flower.

The long thread-like spores first begin to swell at various points, and these give origin to equally slender branchlets, which soon make their way into the tissues at the base of the young flower, by boring through the epidermis. In about ten days or so the mycelium is found to have invaded all the lower parts of the flower, and soon afterwards its fine colourless hyphæ project from the surface, and form complex folds over the young pistil; from the tips of the hyphæ which constitute these folds, hundreds of thousands of extremely minute, colourless conidia are budded off (Fig. 41—*a*), and large quantities of a viscid sugary liquid are excreted: consequently the conidia are bathed in this glairy sweet fluid, oozing out from the flower like drops of pale honey. These drops of sticky fluid have long been well known to farmers all over the world, and many years ago it was the opinion of some that wherever this "honey-dew" (as they called it) was abundant in

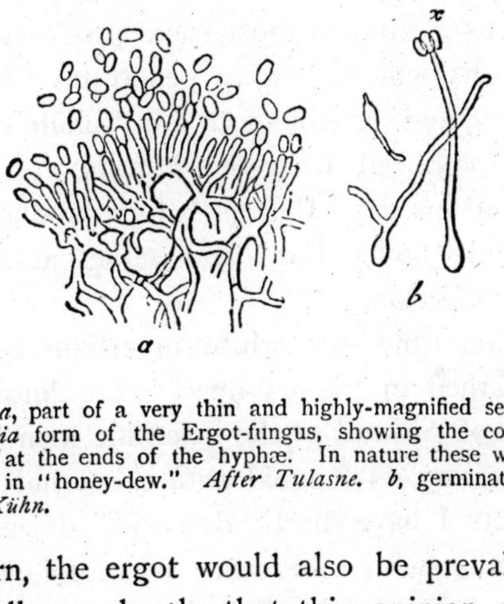

Fig. 41—*a*, part of a very thin and highly-magnified section of the *Sphacelia* form of the Ergot-fungus, showing the conidia being formed at the ends of the hyphæ. In nature these would all be bathed in "honey-dew." *After Tulasne.* *b*, germinating conidia. *After Kühn.*

the corn, the ergot would also be prevalent, and we shall see shortly that this opinion was well founded. But there now comes a discovery which the farmers did not divine. It was found that insects, especially flies, are particularly fond of this "honey-dew," and as they flit about from flower to flower in the corn-fields, they cannot help but carry much of it about, sticking to their straggling feet and inquisitive proboscides; and there can be no doubt whatever that they thus infect healthy flowers in all directions, for, as we shall now see, the conidia forthwith transfer the disease.

When a conidium comes in contact with the base of the young flower, it at once germinates (cf. Fig. 41—*b*), and in most cases produces a germinal hypha which directly penetrates the base of the pistil, and develops into a mycelium exactly like that produced from the *ascospores*. In the course of a few days this, as before, produces the conidia and "honey-dew," and serves as a fresh centre of infection.

For some time the white mycelium and its conidia bathed in "honey-dew" were known by the name of *Sphacelia*,[1] the scientific men of the day recognizing its fungoid nature, but not knowing the facts I have shortly described above as to its connection with *Claviceps*. What follows will show still more clearly that these two supposed genera of fungi are simply forms of one and the same species.

After the mycelium at the base of the pistil

[1] From a Greek word which refers to the shrivelled, gangrenous appearance of the attacked tissues.

has gone on producing conidia and "honey-dew" for many days, the pistil—which would normally have developed into the essential parts of the grain which we use as food—is found to be a mere shrivelled, dead mass of tissue, of no more use to the plant or to man. It will be remembered that the mycelium which gives origin to the conidia and "honey-dew" had invaded the basal parts of the flower and pistil, forming a dense white felt-work in the tissues. This felt-work has now begun to compact itself more and more, increasing at the same time in amount, and its outer cells turn dark-coloured and hard (cf. Fig. 37); it is, in fact, developing into the sub-cylindrical body called ergot—the *sclerotium* of the fungus—and as it grows out from the base, the elongating mass lifts up with it the dead, shrivelled remains of the pistil and the *Sphacelia* which gave rise to the conidia (Fig. 37). These processes of growth and completion of the sclerotium take place as the healthy grains around are filling and ripening, and just before the corn is ready to cut, the ergot sclerotia are mature (Fig. 36), and easily fall to the ground, there to rest through the winter and early spring, ready to repeat their mischief next summer. Of course this is frustrated to a large extent by the rotation of crops now practised in nearly all branches of agriculture, but it is obvious that in grass meadows, and on roadsides, and in patches of sedge, the fungus may go on year after year producing its *Sphacelia* and ergot forms parasitically on the host-plant, and then resting through the winter on the ground, and developing its *Claviceps* stromata or drum-sticks saprophytically on the sclerotium (ergot form).

This phenomenon of periodic desertion of the host is technically called *Lipoxeny*, a word derived from the Greek word meaning abandon. It must be distinguished from the very different phenomenon of *Heterœcism* or *Metœcism*, exhibited by the fungus of the wheat-rust, for instance, where, as we shall see, the one period of life is spent parasitically on one host and the other on another.

It will be interesting biologically to mention another curious fact about ergot: the ripe sclerotium contains, in addition to large quantities of oil, certain poisonous substances which have very remarkable properties. Passing over their uses in medicine, these poisons are found to induce, in those who eat the ergot, or bread, &c. made from grain with which any considerable quantity of ergot was mixed and ground, a very serious malady, which in times of famine during past centuries has been known to take the form of a dreadful epidemic, owing to the prevalence of ergot, and the ignorance and lack of power to choose on the part of those who ate it. This disease manifested itself in the form of a kind of gangrene at the extremities, owing to the paucity of blood supplied by the contracted tissues—the primary effect of the alkaloids in ergot apparently being to contract the muscles, and in fact it is employed for this purpose in obstetric medicine. It is a well-known fact that ergotized grasses are injurious to sheep and cattle, and although there is no proof known to me that the sclerotia can survive the process of passage through the alimentary canal of an animal, it is suggestive that in the case of other plants, and even of bacteria, we know of instances where these poisonous or disease-inducing properties act to the direct advantage of the species by insuring the safety or even the nutrition of the seed or germ. It is not inconceivable that the ergot is advancing in these matters, and improving its advantages in the struggle for existence.

Be this how it may, it will be obvious to all that in the device, so to speak, of *Lipoxeny* the fungus presents an adaptation of great service to the needs of the species, for it ripens its sclerotia at the same time as the grain ripens, and does not produce the spores until the young flowers are developing. Then notice what a subtle trick is played, so to put it, by the presentment of the sweet "honey-dew," and conidia in enormous numbers, to the flies, which rapidly distribute the conidia in all directions.

It is obvious that no method of "dressing" the seed-grain can be resorted to with effect, because the corn is not attacked (so far as we are informed) until it is beginning to flower. Indeed it seems difficult to suggest any feasible preventive measures; but it may be pointed out that the removal of the ergots is possible on a certain scale, and each one removed means so much danger avoided. I question how far the removal of ears which show the extremely infective "honey-dew" and *Sphacelia* stage is feasible. Probably the most practicable method is to harvest the crop as soon as possible, and thus take away the ergots before they fall to the ground. Specifics of all kinds may safely be regarded as useless in the present state of our knowledge.

CHAPTER X.

THE HOP DISEASE.

THIS malady, known only too well to the hop-growers of Kent, is due to the ravages of a fungus belonging to the group of true mildews, the *Erysipheæ* of mycologists; the word mildew comes from the German *Mehl-thau*, and refers to the floury, dusty appearance on the leaves affected by these pests. Among numerous other known forms of these mildew fungi, there is one[1] that attacks very many of our wild and cultivated plants, such as the dandelion, teazel, plantain, veronica, balsam, cucumbers, &c., and which has made itself especially obnoxious by flourishing on the hops. We have here, in fact, a particularly instructive example of a disease which became suddenly interesting as soon as it touched our

[1] The hop-fungus is *Podosphæra Castagnei*, very common on many plants in Europe and elsewhere.

pockets, although few would concern themselves with it so long as it confined its ravages to wild plants and weeds; whereas more can be learnt about it on such plants as the weeds above mentioned than on the hop, which happens to be a somewhat unfavourable plant for investigation. However, this and other objections soon disappeared when attention was once directed to the matter, and we probably know as much about the hop-mildew as we do of any parasitic fungus, and the story is as interesting as it is important.

It appears that the hop-mildew has been well known for many years, but it is only occasionally that the pest gets the upper hand, owing to favouring conditions of the climate at the period when the young hops are showing, and in particularly bad wet seasons it has been known to destroy the crop so far that the culture has had to be abandoned.

The symptoms of the disease are the appearance of pale spots on both sides of the leaves; these spots become white and dusty-looking, as if covered with flour, increasing in area at the same time, and especially on the lower surface of the leaf, where the epidermis is thinner, more sheltered, and consequently affords more moisture.

As time passes, the white "mildew" spreads on to the young hops, and this is the symptom so dreaded by the hop-grower. In this stage it is very like the so-called vine-mildew to the unaided eye (cf. Fig. 2), whence the confusion between them.

After some weeks, the patches assume a darker colour, especially those on the lower sides of the leaves and scales, and eventually the larger patches become nearly black. These changes in hue are almost entirely due to the gradual changes in colour of the fungus as it grows older. It is at first pure white, and gives off multitudes of white conidia (whence the floury appearance of the younger spots), and then darkens to brown, and finally develops numerous black-brown fruit-bodies.

If the lower surface of a leaf is examined with the microscope, selecting a part affected with the white "mildew," it will be found that the epidermis is covered with an interlacing network of colourless, septate, branched hyphæ, running over the surface, and crossing one another in all directions. From these prostrate filaments running on the surface, there may be developed erect branches which stand off vertically into the air, and which behave in a very curious manner, as we shall see directly, and give origin to colourless, glistening conidia. The whiteness of the young disease-spot is due to reflection of the light from these colourless fila-

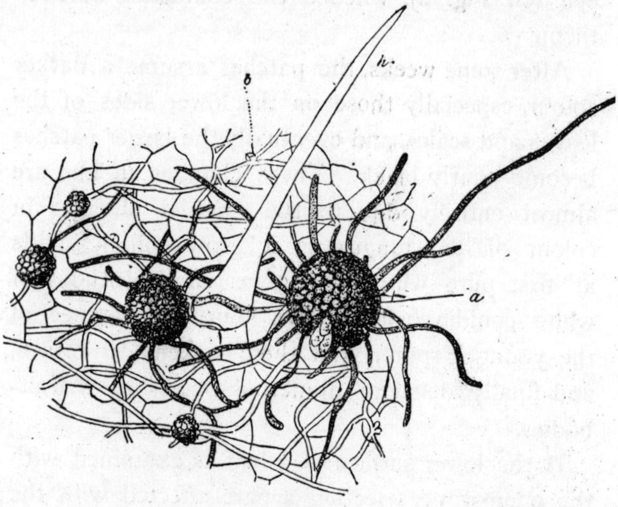

Fig. 42—Piece of epidermis of hop, showing mycelium (*b*) and perithecia (*a*) of the hop-mildew on its surface. At *h* is a large hair. At *b* the first beginnings of a perithecium are appearing. Magnified.

ments, and the "floury" appearance is intensified by the myriads of dust-like conidia given off.

Closer examination of the prostrate hyphæ shows that each is a delicate tube of cellulose, enclosing water and protoplasm, and divided into cylindrical joints by cross-walls; moreover, it is not difficult to observe that the filament is not merely lying on the epidermis of the hop-leaf, but is closely fixed to it at various points, so that if one could take a given branch by the end, it would be impossible to remove it without rupture at the fixed points—much as it would be impossible thus to remove a length of leaden gas-piping, as usually fitted in houses, without tearing out the hooks which hold it to the wall. In the case of the fungus, this is because the hyphæ send little boring organs, called *haustoria*, through the cell-walls of the epidermis, and these *haustoria* serve the double purpose of nailing the tubes to the leaf, as it were, and of absorbing food-materials from the cells; whence the name *haustorium*, or absorbing organ (Fig. 43).

The branches which stand vertically off from the hyphæ just examined, are similarly constituted as regards their being cellulose tubes containing protoplasm, but the joints or segments are short and stout, and soon break off from one another. It is these broken-off segments which act as the conidia, and intensify the meal-like appearance of the fungus. It is only in certain seasons, when the air is damp for some time, that these conidial segments are developed in any quantity, but when such is the case they are so numerous, and so easily carried from plant to plant, that no wonder need

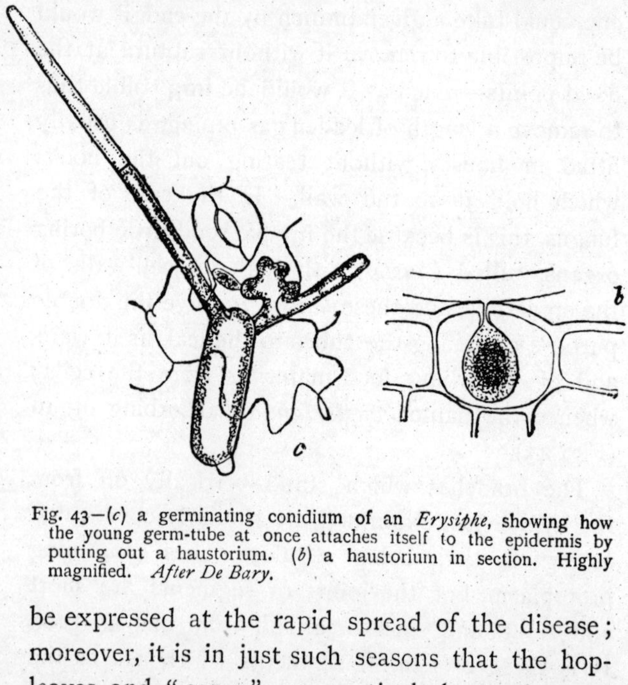

Fig. 43—(*c*) a germinating conidium of an *Erysiphe*, showing how the young germ-tube at once attaches itself to the epidermis by putting out a haustorium. (*b*) a haustorium in section. Highly magnified. *After De Bary.*

be expressed at the rapid spread of the disease; moreover, it is in just such seasons that the hop-leaves and "cones" are particularly tender and watery, their cell-walls thin, and their recuperative power low, whence we have other important factors favourable to the spread of the fungus. Finally, it is in such wet surroundings that the conidia meet with the best conditions for rapid germination.

The conidium, after separation from the supporting hypha, is a somewhat ellipsoidal or barrel-shaped body, consisting of a quantity of watery protoplasm surrounded by a delicate, colourless cell-wall. On germination, one or more hyphæ are developed from the ends of the conidium (Fig. 43); and if the process is going on in a drop of water on the surface of a leaf, the hyphæ go on growing and branching on the epidermis, and here and there put forth haustoria through the walls of epidermal cells at the points of contact (Fig. 43). In this way is gradually developed the web-like mycelium which forms the white spots and patches; and it is important to notice that we have here a type of parasite very different from those hitherto examined, in that it never enters the plant, either by sending its hyphæ through the stomata, or by piercing the tissues, further than to send its tiny haustoria into the cells of the epidermis. This fungus is, in fact, a parasite which lives *on* the host-plant, simply tapping it at various points by the haustoria, which also serve for holdfasts. It is termed, therefore, an *epiphytic* parasite, in contradistinction to parasites which live *in* the tissues of the host, and which are called *endophytes*.

It will be noticed that as the patches of mildew increase in size, &c. with age, the white tinge is gradually replaced by a yellowish one, and this in turn by a brown or nearly black colour. Even a lens shows that the dark colour is partly due to a number of small, spherical, brown bodies scattered like minute beads on the mycelium; and the microscope shows that there are series of these tiny structures, beginning with mere yellowish points, and ending with brown or black spheres, each as big as a small pin-head (Fig. 42). These bodies are the *perithecia*, or true spore-cases[1] of the fungus, each of them enclosing a number of spores which serve to reproduce the parasite in the following spring. Each ripe *perithecium* is a tiny spherical box, consisting of a somewhat brittle shell fixed to the hyphæ of the mycelium at its base, and enclosing a pear-shaped and very delicate bag, called an *ascus*,[2] filled with spores.

[1] Therefore called *Sporocarps*: also termed *Cleistocarps*, owing to their having no orifice, but being entirely closed in.

[2] In this particular, this *Podosphæra* differs from other Erysipheæ, which have *several* asci in each sporocarp.

We will examine each structure a little more in detail.

The dark brown shell is bright outside, as if polished, and is divided up into numerous five or six-sided areas (the outer walls of cells composing it). Here and there it gives off long, brown, tubular hyphæ, which radiate on to the mycelium and epidermis below, and help to fix the whole (Fig. 42). This spherical shell is attached at one point below, as said, to the mycelium (Fig. 44).

Inside the shell, the pear-shaped *ascus* arises from the base, being attached by its narrow end to that portion of the *perithecium* where it rests on the mycelium ; in the ripe *ascus* are eight ovoid spores, called *ascospores*. It is possible to see most of the stages of development of these curious structures, by proper examination with the microscope, and since the process is typical and has been well investigated, I give a short sketch of the chief phenomena.

At certain places where the branched and septate hyphæ of the mycelium cross or touch one another on the epidermis, two short branches are put out into the air (Fig. 44—*A*). As these erect branches grow, one is found to bend close to the other, and remain in contact with it.

Then a number of short branches spring up from the base of the two just mentioned, and we have thus a sort of tuft of short hyphæ all ascending into the air from one point (Fig. 44—*B* and *G*). I shall abstain from discussing certain theoretical matters, which necessitate the giving of definite

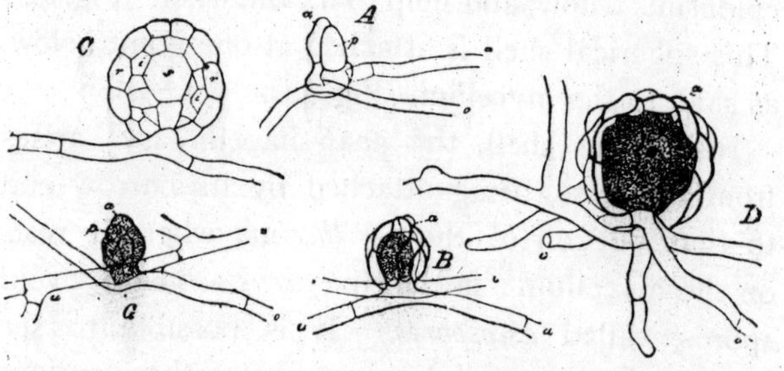

Fig. 44—Various stages in the development of the sporocarp of the hop-mildew (*Podosphæra Castagnei*), showing the contact of the two short branches (*A*), one of which (*p*) gradually becomes invested by enveloping branches (*B*). The envelope then forms the outer wall of the sporocarp, and the ascus is formed from the enclosed branch (*C* and *D*). Highly magnified. *After De Bary*.

names to each of the two branches first mentioned, as well as those which arise from the basal parts, simply pointing out that one of the two first-named short branches comes to be in the middle, and the rest grow up close around it, so that the little tuft now consists of about seven short hyphal branches—one in the centre, and about six closely surrounding it, and covering it in at the sides and top (Fig. 44—*B*). It must suffice for our present purpose to show that the central one becomes the *ascus*, while the enveloping branches develop into the shell (Fig. 44—*C* and *D*). The young *ascus* is filled with protoplasm, which soon breaks up into eight portions, each of which becomes a spore. The shell is completed by the formation of septa in the enveloping branches, and by the walls turning dark brown as they grow older, and it is these septa which give the areolated appearance to the outside of the shell

When the *perithecia* have been formed and ripened in large numbers, especially on the shaded lower sides of the leaves, the mycelium is practically exhausted, and as these processes are completed towards the end of the summer, the leaf so onfalls. The hard shell insures the protection of the *ascus* and its spores during the winter, and in the following spring the *ascospores* germinate and start the fungus once more. At first, only one or two of the white mildew specks appear; then it spreads from leaf to leaf, and soon a whole field of hops may be covered.

It is obvious on a little reflection that the direct injury done by this fungus is not likely to be very great unless the mycelium is present in large quantities; for such injury consists, first, in the abstraction of substance from the epidermis cells by means of the almost immeasurably tiny haustoria, and, secondly, in the obstruction to light, air, &c., which the mycelium interposes between the environment and the leaves or young hops. In ordinary fairly dry seasons, the hop or other host-plant may be perfectly able to pay the tax which the fungus imposes on it; but it is in the wet and unfavourable seasons that the hop especially suffers, and it must not be forgotten that such seasons would be bad for the cultivator even if the fungus were absent—much more so, therefore, when the pest is rendered more abundant and flourishing; for it is just such seasons which favour the fungus at the expense of the host.

A word or two may be added as to the employment of remedial measures. There is no question that the sulphureous gases, such as are evolved from oxidizing powdered sulphur, for instance, kill the delicate mycelium immediately they dissolve in the water of the hyphæ, and reach the protoplasm; and a good deal has been done by dusting the hops with finely powdered sulphur, or a mixture of sulphur and lime, blown on by means of mechanical bellows of various kinds. In quantities, and if in contact with the leaves for some time, the sulphureous gases may damage them irretrievably, and it is questionable if the tissues do not always suffer; but the difficulty here is to effect a compromise, and choose the less of two evils, and that has been done by checking the ravages of the fungus with sulphur. At the same time, it should be clearly borne in mind that the *Podosphæra* of the hop-disease is a common parasite on numerous well-known weeds of the fields and road-sides, as already stated, and thus clean cultivation may come to mean a great deal to those who have such neighbours.

CHAPTER XI.

THE "RUST" OF WHEAT.

FEW parasitic fungi are better known than those to which the pest which causes the "rust" of wheat belongs, and no group of organisms has proved of more importance to the biologist than these, not only because his study of them has disclosed so much concerning parasitism in general, but also on account of their peculiar transformations. To the farmer they have always been of only too much importance, because few parasites are more difficult to contend against when once they obtain a firm hold on the crops. Fortunately, however, the results of investigation on all hands have been put into practice with such effect, that although the disease is still apt to do much damage, it is comparatively rare to find it devastating whole districts in the way it did formerly.

During the months of June and July the bright fresh green leaves of the wheat are occasionally seen to lose colour day by day, and the observer detects numerous orange-yellow lines and patches on the surface of the foliage (Fig. 45—*u*); these elongated patches become larger and more numerous day by day, emitting countless bright orange powdery granules if rubbed or shaken, and the malady may often be seen spreading from one plant to another as time goes on. It is in great part due to the yellow hue of this powder that the paler look of premature ripeness comes over the patch of wheat.

At a very early period in the study of fungoid pests, this yellow powder was found to consist of spores—each yellow grain being a single spore —and the fungus in this state was named *Uredo*,[1] and was described and figured as a definite genus of fungi. The bright orange-yellow spores have hence received and retained the name of *Uredo-spores.*

If one of the affected wheat-plants is marked and carefully watched, it will be found that in the course of three or four weeks the linear patches do not increase, but they become purple-brown or

[1] From a word meaning to burn or shrivel up.

almost black in colour, and the powder which

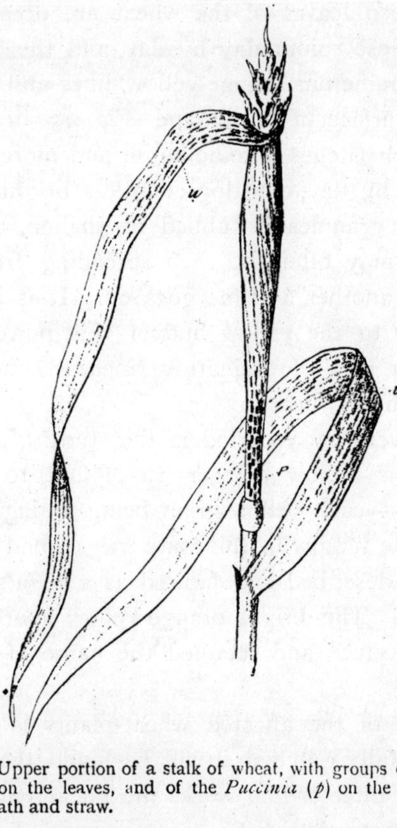

Fig. 45—Upper portion of a stalk of wheat, with groups of the *Uredo* (*u--u*) on the leaves, and of the *Puccinia* (*p*) on the fast-ripening leaf-sheath and straw.

escapes from them has changed its characters,

and especially its colour. The grains (spores) are no longer bright yellow, but dark purple-brown (Fig. 45—*p*). It can be observed that these dark spores are at first only produced in sufficient quantity to give a dirty hue to the yellow masses of Uredospores, then they predominate more and more, until by the time the corn is fully ripe and the straw has assumed its pale characteristic mature colour, the linear spots are found to yield only these dark purple-brown spores. In years gone by people were much puzzled by these transformations, but they described the dark patches on the ripening straw as "mildew," and botanists gave the fungus the name *Puccinia*, in honour of an Italian named Puccini. It was found that the dark-coloured spores are not only produced later in the season than the Uredospores, but they also differ from them in size, shape, markings, &c., as well as in colour, and thus observers were quite justified in regarding the two forms as separate fungi. Nevertheless, the fact of careful observation impressed itself more and more on people—the *Uredo* not only preceded, but often seemed to turn into the *Puccinia;* and many held that the "rust" (*Uredo*) and the "mildew" (*Puccinia*) were in some mysterious manner connected.

As we shall see, this idea of connection between the two forms is quite correct, the so-called *Puccinia* being only an autumnal form of the *Uredo*. But before proceeding to the establishment of this fact, it will conduce to completeness if we first notice another opinion which prevailed at various times and in different localties.

It was often noticed and insisted upon that the neighbourhood of bushes of the common barberry was injurious to the wheat. No very definite ideas or reasons for the belief were always forthcoming, but farmers and others maintained more or less doggedly that the propinquity of the barberry favoured the development of "rust" on the wheat. Several able observers upheld this view, while others denied that there was any foundation for it beyond mere superstition. Nevertheless, a law was passed more than a hundred years ago calling on the farmers of the state of Massachusetts to destroy all the barberry bushes, and the Norfolk agriculturists of the same period firmly maintained that the barberry blighted the wheat.

At the beginning of the present century Sir Joseph Banks suggested it as probable that a certain bright orange-yellow fungus common on the barberry (Fig. 46) may be the same as that causing the rust of wheat; and a Danish observer of the same period convinced himself that the fungi are connected, but his methods were not conclusive to others at the time.

One great difficulty that had to be overcome

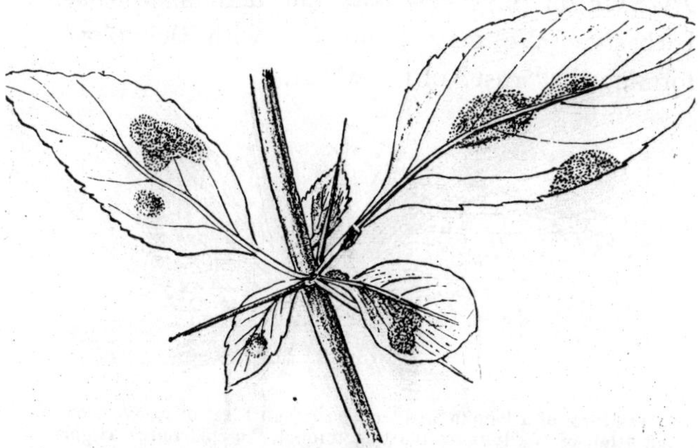

Fig. 46—Portion of a branch of barberry, with a tuft of leaves attacked by *Æcidium Berberidis*, which forms yellow cushions on the leaf-blades and stalks.

before these suggestions could be received with favour was this. Examination by the botanists and with the microscope showed that the fungus on the barberry presented hardly any resemblances to the *Uredo* on wheat beyond the resemblance in colour. The fungus on the barberry received the name of *Æcidium*,[1] and was known for many years as a separate genus.

Before going further, it will be well to look at the structure and other peculiarities of these three forms of fungus—the so-called *Uredo* and the *Puccinia* on the wheat, and the *Æcidium* on the barberry. We may commence with the *Uredo* forming the "rust" of the wheat.

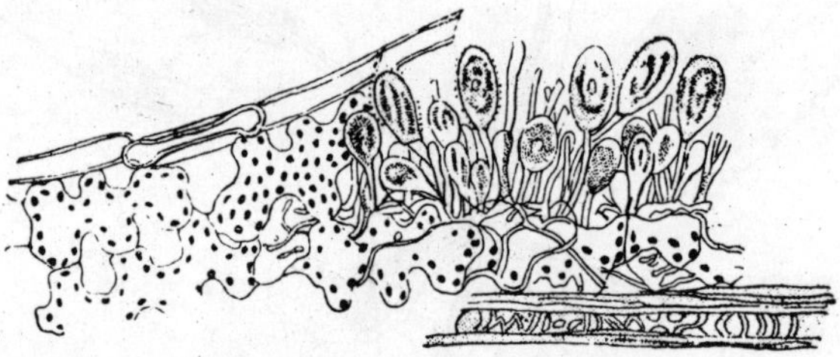

Fig. 47—Piece of a longitudinal section through a leaf of wheat, showing a tuft of Uredospores bursting through the epidermis. Highly magnified.

If one of the leaves is cut across, so that the section passes through one of the linear orange-yellow patches, it will be found that the patch is a rupture in the tissues of the leaf, due to crowds of the yellow spores (Uredospores) having been formed beneath the skin of the leaf until they burst through to the outside (Fig. 47); the torn edges of

[1] From the Greek αἰκία, referring to its injurious character.

the skin are easily seen in a suitable preparation. Deeper down the tissues of the leaf are found to have fungus hyphæ running irregularly between the green cells, and these hyphæ are branched, septate, and often contain small orange-coloured drops in their protoplasmic contents. Comparative studies have shown that these hyphæ derive their nourishment from the contents of the green cells, and therefore exhaust the leaves and plant of valuable and hardly-earned food-substances which should have gone to form grain, and that the ends of branches of the hyphæ eventually turn towards the surface of the leaf, blocking up the spaces just below the *stomata*, and finally bud off the bright orange-yellow Uredospores from their tips (Fig. 47). As the spore-bearing branches increase in number and the separated spores accumulate, the pressure bursts the skin of the leaf, and we have the linear patches covered with their bright yellow dust-like spores.

Placed under a high magnifying power, each of these Uredospores is a somewhat oblong or egg-shaped cell, with a not very thick coat studded with short spicules, and containing protoplasm in which yellow oily drops abound (Fig. 48). Their average size may be taken as one thousandth of an inch long. Each spore becomes detached by breaking away from the branch on which it was produced, and is seen to have a few thin round spots like pores at its equator.

When these Uredospores, fresh from the wheat,

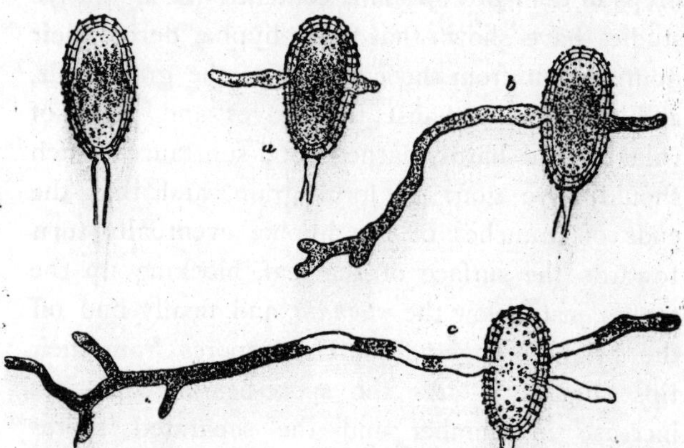

Fig. 48—Germinating Uredospores of *Puccinia Graminis*, showing the various stages of development of the germ-tubes, *a*, *b*, and *c*. Very highly magnified.

are put into water, they soon germinate, sending out two or more thin tubes from the above-named thin spot (Fig. 48); these tubes are composed of a delicate wall of transparent cellulose, containing the watery protoplasm and orange-coloured, oil-like granules and drops previously seen in the spore itself. The tubes soon become unequal, one of them growing faster and developing in a few hours into a delicate, sinuous, and branched filament some ten or twelve times as long as the spore (Fig. 48—*b* and *c*). If it remains in water

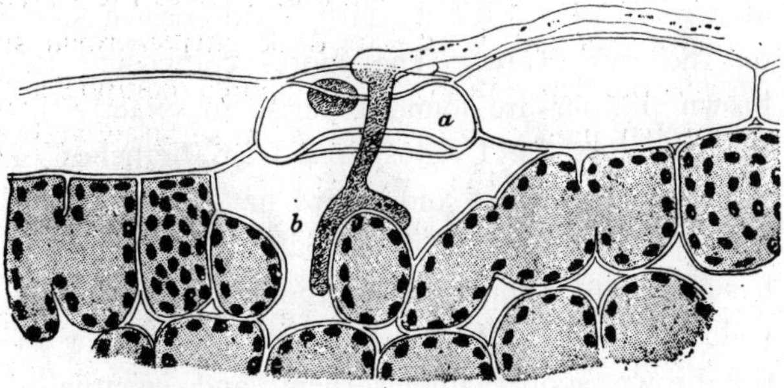

Fig. 49—Portion of longitudinal section of a leaf of the wheat, showing germ-tube of the *Uredo* passing through a stoma (*a*) into the intercellular space (*b*). The letter *a* is placed in a guard-cell of the stoma; the other guard-cell having been removed. Very highly magnified.

the whole soon perishes; it only lives so long as the substances contained in the spore last out, but when they are exhausted it dies.

It can easily be proved that if such a Uredospore is allowed to germinate on a young fresh leaf of a living wheat-plant, the tip of the germinal tube finds its way into one of the *stomata* (Fig. 49);

M

enters into the tissues of the leaf, grows up into a new mycelium spreading between the cells of the leaf, and finally—*i.e.* after about a fortnight to three weeks—again develops Uredospores. The *Uredo*, therefore, is capable of reproducing the *Uredo* on the wheat.

We will now turn our attention to the *Puccinia*, as it was called, with its dark purple-brown spots on the ripe straw. The spores composing the brown powder are found to arise in exactly the same way as the Uredospores, from branches of the same mycelium, and, as we have seen, at the same places (Fig. 45); in fact, after the mycelium has gone on for some weeks producing the golden-yellow Uredospores, it begins to form a few of the brown spores among them, and eventually ceases to produce any other kind but these dark brown spores, and then the spots turn deep coloured, almost black (Fig. 45—*p*). The consequence was, as soon as it was firmly established that these dark late autumn spores are developed from the same mycelium as the yellow summer spores (Uredospores), the name Teleutospores was applied to the former, the term signifying spores formed late or last in the life-history of the fungus.

Any one of these Teleutospores examined as before under the microscope is found to differ from the Uredospore in being longer and more spindle-shaped or club-like, though like them they are borne on stalks (Fig. 50): moreover, the long spore is divided by a partition across the middle, so that

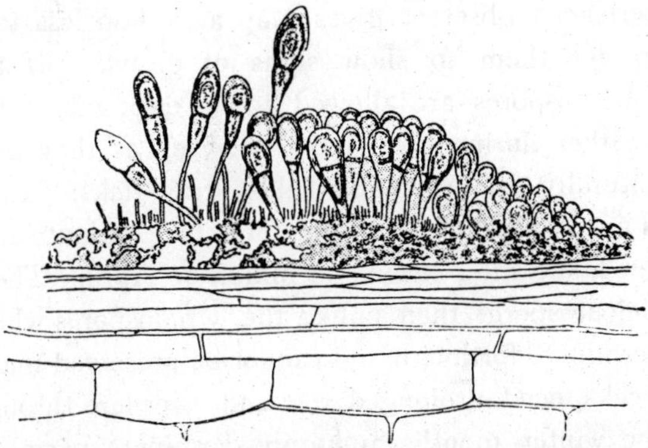

Fig. 50—Longitudinal section through a patch of the Teleutospores of *Puccinia Graminis* on a stalk of wheat-straw. Highly magnified.

it consists of two cells or chambers instead of one only. In addition to the brown colour, the coat of the Teleutospore is also much thicker and harder than that of the Uredospore, so that, all things considered, there are several obvious points of distinction which would prevent any observer from mistaking the two kinds of spores (cf. Figs. 47 and 50).

But the differences are much more emphasized when we sow the fresh Teleutospores in water as before. Instead of rapidly germinating in a simple manner as did the Uredospores, they lie unchanged for days, weeks, and even months, until the inexperienced observer gives it up as a hopeless task to get them to show signs of growth. If the Teleutospores are allowed to lie exposed to the weather during the winter, however, as they may naturally do in stubble-fields, straw-ricks, &c., it is not at all difficult to get them to germinate in a few hours after sowing in March or April. These Teleutospores then, unlike the Uredospores which germinate forthwith and cannot be preserved many weeks, need prolonged rest and exposure through the winter months, and only germinate normally in the spring after they are developed. I may add that they may be preserved alive much longer than this, and even after three years some of them may be caused to germinate after a few days in water.

The products of the germination are conspicuously different from what we have seen before. In principle, the process commences as in previous examples given, but each of the two chambers of the Teleutospore puts forth (through a thin spot in its coat) a delicate tube, which looks at first as if

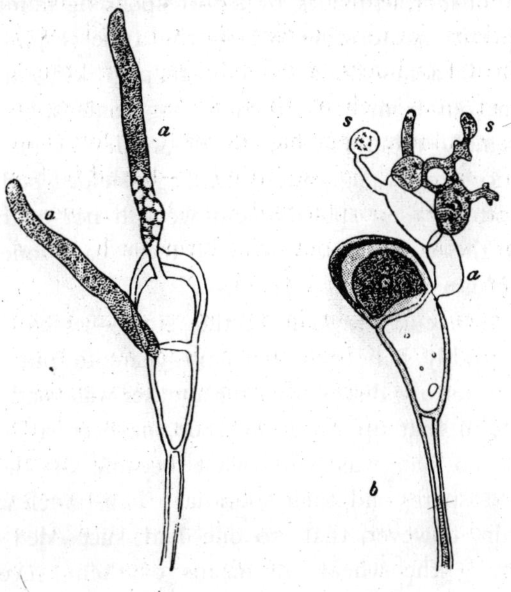

Fig. 51—Two Teleutospores of *Puccinia* germinating. In the one to the left each cell has given off a promycelium (*a a*); in *b* (to the right) only the lower cell has done so, and the promycelium (*a*) has given rise to sterigmata bearing sporidia (*s s*). Very highly magnified.

it was about to grow on as a simple tube like those examined previously. Not so, however, for when the tube from either chamber has reached a length of about two or three times that of the spore, it ceases to grow longer, and becomes jointed into three or four segments, by a corresponding number of partitions forming across the tube (Fig. 53). Then, in a few hours, a delicate peg-like branch is put out from each of these segments, and an extremely minute, egg-shaped swelling forms at the tip of each peg; these little egg-shaped bodies are often called sporidia, and we will not quarrel with the word here, but will term each one a sporidium as usual (Fig. 51—*s*).

Now a curious point in this story—which you will readily believe took many years to put together, as one discovery after another was made—arose, in that up to about 1860, or even a little later, no one was sure what became of these Teleutospores and their sporidia. Thus much was certain, however, that no one had succeeded in infecting the wheat by means of them; every attempt to produce the "rust" or "mildew" on green wheat or older leaves by means of sowing the Teleutospores and the sporidia utterly failed.

Things had been in nearly this condition for a long time when the late Prof. De Bary took up the matter, and his attention was turned to the old story which connected the barberry with the wheat-rust. De Bary made a very accurate study of the fungus on the leaves of the barberry—the so-called *Æcidium*—and thoroughly sifted the question as to any possible connection between it and the *Puccinia* on wheat.

If the yellow rounded patches on the barberry leaves in spring are carefully examined with a lens, they present several features of note. In the first place, each bright yellow patch (Fig. 46) is thicker than the rest of the leaf, forming a sort of waxy cushion; secondly, one observes a number of minute pore-like dots on the upper side of these cushions (Fig. 52—*s*), and certain cup-like depressions on the lower side (Fig. 52—*æ*), from which a golden-yellow powder escapes. These depressions on the lower surface are the cups or *Æcidia*, and the yellow powder consists of the spores of this fungus; and since these spores are developed in the *Æcidium* cups, they are called *Æcidiospores*.

The structure and mutual relations of all these bodies are best made out by cutting vertical sections through the leaf, and especially, of course, the cushion-like swelling (Fig. 52). The cups are then found to arise as balls of fungus hyphæ, springing from the mycelium in the leaf, just beneath the epidermis or skin of the latter. Each of these balls contains, at the part attached to the fungus mycelium, a series of hyphæ, which are short, and have their ends directed in a radiate manner towards the opposite pole of the ball, that is, the part nearest the epidermis; the remainder of the cavity of the ball is filled with closely-packed

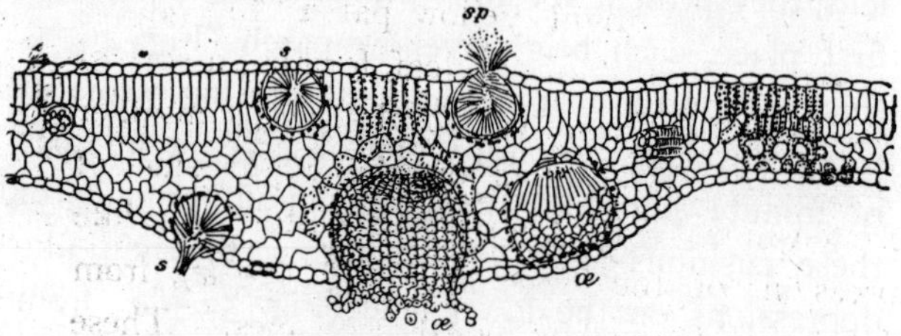

Fig. 52—Vertical section through a patch of Æcidia (*æ*) and Spermogonia (*s*) on the barberry leaf, showing the swollen tissue at the diseased part. The small æcidium to the right has not yet burst through the epidermis; at *sp* the spermogonium is extruding spermatia. Highly magnified.

spores, so like the Uredospores already described, that they may be said to differ only in their more equally rounded shape, often rendered polygonal by compression. Soon after this stage has been reached, the growing ball bursts the skin of the leaf, and projects to the exterior; then it opens at the apex, and the edges of the ruptured coat turn back, and the structure now presents the appearance of a sort of cup, or deep bowl, from which the yellow rounded *Æcidiospores* are escaping, and new ones being developed from the hyphæ at its base (Fig. 52—*a*).

It is not necessary here to go further into the details of the structure of these cups, nor need we dwell upon the curious little bodies (called *Spermogonia*) found on the upper side of the cushion (Fig. 52—*s*); it suffices to say of the latter that they are smaller hollow balls, the hyphæ in which sprout off certain extremely minute spore-like bodies, but are of no importance to our description. I am not saying that they have no bearing upon the matter, but simply that they need further investigation, and do not concern us here.

Now De Bary made the astounding discovery that if the Teleutospores of the *Puccinia*, which have remained on the wheat-straw through the winter, are sown in spring on the young barberry-leaves, they germinate and produce the sporidia as described, and these latter infect the barberry-leaf by sending delicate germinal tubes of protoplasm directly into the cells of the leaf (Fig. 53). Then, in a fortnight or three weeks, the mycelium begins to so alter the leaf at the place affected by attracting so much food-material to that neighbourhood, that the cells are over-fed, enlarged, and consequently thicken the leaf-tissue to form the cushion-like swelling above described; and on this cushion arise the *Spermogonia* and *Æcidium* cups already described (Fig. 52).

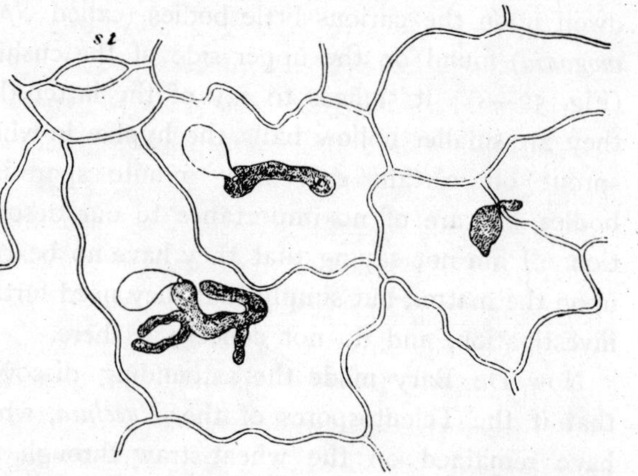

Fig. 53—Sporidia of *Puccinia Graminis* germinating on the epidermis of a barberry leaf, and putting out germ-tubes which bore through the cell-walls. Very highly magnified.

When, in 1864, it was first announced that the Teleutospores of the wheat *Puccinia* had thus been proved to give rise to the *Æcidium* of the barberry, the statement was received with quiet reserve by a few who knew and preferred to wait, with doubts and smiles by many more, and with open disbelief by others. But the discoverer of this remarkable phenomenon—the power of a fungus to live in different forms on two or more distinct hosts, and hence called *Heterœcism*[1]—was too sound a man to leave any gaps in the evidence; and in 1865 he proved that not only does the *Puccinia* of the wheat give rise to the *Æcidium* on the barberry, but the *Æcidiospores* of the latter, if sown on the young leaves of corn, infect them and produce the *Uredo* and then the *Puccinia* on them.

Probably no subject in the whole range of Biology has given rise to more discussion, and even consternation, than these phenomena; for, be it noted, it is not only that the fungus in question lives at different seasons on different host-plants, but it has a special form or forms on each. We have got used to these phenomena now, however, since we know of many other similar cases, and no generalization is better established than those of the polymorphism and heterœcism of these fungi. To mention one or two cases only, a fungus known as *Gymnosporangium* is parasitic on the juniper, but its spores need to be sown on the hawthorn,

[1] From two Greek words meaning different dwellings; also called *Metœcism*.

and then produce a form (utterly different from the *Gymnosporangium*) known as *Rœstelia*. The spores of the *Rœstelia*, again, need to be sown on junipers, and then give rise to the *Gymnosporangium* as before. Again, it took a good many years of sharp observing and careful experiment before it was found out that a certain fungus long known as *Coleosporium*, and parasitic on that common weed the groundsel, is only one form of the fungus called *Peridermium*, which does so much damage to pines. The form on the groundsel is very different from that on the pine, and it cannot be wondered at that they received different names, and we are now being educated by experience—and some of it has been very bitter experience—to recognize that if we would not be utterly swamped by these insidious enemies to our crops and supplies of various kinds, it will be necessary to extend this elementary knowledge of the subject.

The discoveries sketched in this chapter are already bearing fruit. Few doubt that wheat-rust is more rigorously kept in check now that botanists have explained the nature of the disease, and the kinds of danger to be prepared against, and the apathy displayed with regard to these matters is probably to be referred, not to any real want of interest on the part of those concerned, so much as to ignorance of what has already been done even in this infant stage of pathology, and the scientific agriculture and horticulture of the future, as well as forestry and the planting industry generally, will have to concern themselves with these matters much more liberally than they have hitherto been moved to do.

CHAPTER XII.

CONCLUSION.

IN bringing this little book to a conclusion, it is to be remarked that although types of the chief diseases of plants, so far as they are caused by parasitic fungi, have been dealt with, it is by no means to be supposed that the subject has been or could be fully treated in so short a space. It may be well, therefore, to give a brief reference to some headings left untouched, in order that the reader may form an idea of the further ground to be traversed from the various starting-points given.

In the first place there is an enormous group of parasitic fungi related to the *Pythium* which we studied in Chapter III., and the *Phytophthora* of the potato-disease (Chapter V.). This group is called the *Peronosporeæ*, from the name of one of its principal genera—*Peronospora*. The fungi of this alliance cause many of the worst diseases known in agriculture, &c. To give a few instances only, in addition to those already examined: species of *Pythium* destroy numerous plants, and seedlings are killed by species of *Phytophthora* as well; and many diseases, such as those of the onion, beet, spinach, poppy, and many garden plants are due to *Peronospora*. *Cystopus* destroys cress, radishes, turnips, and many other garden plants. The salmon-disease is associated with a species of *Saprolegnia*, and numerous allied forms destroy insects, algæ, &c. Even some of the *Mucor* family are parasitic on fungi.

The curious *Plasmodiophora*, which formed the subject for Chapter IV., occupies an isolated position, as a parasitic and anomalous member of a great group of organisms known as the *Myxomycetes*, and it must therefore be regarded as standing alone; it certainly has nothing directly in common with any of the other types.

The *Ustilago*, causing the "smut" in corn, serves as the nucleus of another grand division of the fungi, and we may regard the *Ustilagineæ* as centering more or less around it. Beginning with *Protomyces* and *Entyloma*, simpler forms which injure certain weeds, such as the dandelion, goutweed, figwort, poppy, &c., the student should study the genus *Tilletia*, which causes the "bunt" of wheat and many other grasses, and species of which attack various *Polygonaceæ* and other plants. The genus *Schrœteria* injures veronicas; *Doassansia* lives in water-plants; *Tubercinia*, *Sorosporium*, *Thecaphora*, and others, inhabit various herbs; and *Graphiola* is found on palms. Even among these exquisitely parasitic forms, the *Ustilagineæ*, however, we meet here and there with species which do little or no harm to the host, and recent researches make it not improbable that some of them are actually beneficial to the latter, living symbiotically with it, and possibly doing work for it.

The fungi dealt with in Chapters VII., VIII., IX., and X., supply us with another rallying-point for our ideas, and they all agree in one remarkable and important feature, the production of *asci* containing *ascospores*, whence these types are usually combined (with numerous sub-types) in a grand circle of alliance named the *Ascomycetes*. Convenient as this view is, however, it is well to bear in mind that marked differences in all minor characters are observable between the groups of fungi most closely allied to the *Exoascus* of "bladder plums," the *Peziza* of the lily-disease, the ergot of rye, and the hop-mildew respectively; and I shall therefore regard each of these types as the central point of certain groups of fungi.

Allied to the *Exoascus* which causes "bladder plums," are a number of forms which injure the leaves of peaches, apricots, &c., and which produce deformities in the branches or leaves of alders, birches, poplars, and other trees. The peach "curl," caused by *Exoascus deformans*, for instance, is well-known in this country, and the carpels of willows and poplars are often disfigured by *E. aureus*.

The *Peziza* of the lily-disease, and its allies, serve as an excellent type of the destructive forms of *Sclerotinia* which ruin clover-fields, and damage hemp and onions, snowdrops, carrots, turnips, and numerous other cultivated plants. And the complete *Peziza* is so like the form which does so much injury in the so-called "larch-disease," that it is usual to associate them closely. Less obviously related to these typical *Discomycetes* are the genera *Phacidium*, *Rhytisma*, *Hysterium*, &c., several of which do considerable injury to trees by destroying or interfering with the normal action of their leaves.

The *Claviceps* which causes ergot must be taken

as the type of an enormous group of fungi known as the *Pyrenomycetes*, though if my object had been more specially to set forth the details of classification of fungi, I should have added several simpler forms to this more complex one. The simpler allies in question are found all over the world producing black and brown spots on leaves, &c., and although I do not wish to be understood to say that all small black spots on leaves are due to these fungi, yet it is remarkable how many of these common disfigurations are caused by parasites of this group. As a few examples only I select the following. Hyacinths and many other plants are injured by *Pleospora*; willows, &c., by *Capnodium*; conifers by *Trichosphæria* and *Herpotrichia*; oaks, potatoes, &c., by *Rosellinia*; laburnums and other trees by *Cucurbitaria*; strawberries, mulberries, and many other plants have their leaves spotted by *Sphærella*; the common spots on apples and pears are caused by *Stigmatea* and allied forms; and a very long list of these spot-causing parasites could be drawn up if necessary.

Somewhat more complex, but in the same circle of alliance, are the species of *Nectria* which produce cankers on beeches, apples, currants, &c.; the *Polystigma* which forms orange-scarlet patches on the leaves of plums; and the curious *Epichloe* which does so much damage to the common rye-grass of pasture meadows. Here also comes our selected type, the *Claviceps* of ergot.

Passing on now to a further group, which may be regarded as entirely distinct from any of the preceding, we have the important group of the *Uredineæ*, excellently typified by the fungus of the "rust" of wheat, studied in Chapter XI. Here we meet with parasites which, like *Uromyces* and *Puccinia*, live on the leaves and stems of herbaceous plants, injuring our beans, peas, clovers, beet, asparagus, mint, cereals, and numerous gardeners' plants and grasses; *Phragmidium* and its allies attack roses, raspberries, and their relations; *Gymnosporangium* injures junipers and fruit-trees; *Melampsora* attacks conifers, poplars, and other plants; *Coleosporium* is a great enemy to foresters, as its alternate form destroys the leaders, &c. of pines; and allied plants are injured by *Chrysomyxa.*

It remains to add that several forms of what are popularly termed Toad-stools, belonging to the genera *Agaricus*, *Polyporus*, *Hydnum*, *Stereum*, &c., also do enormous injury to trees and timber; and that these constitute yet another group of fungi. I have not treated of this type in the present work, however, because, in the first place, they are of more special importance to foresters than to farmers and gardeners, and, in the second place, they occupy a large share of attention in my book on *Timber, and some of its Diseases*, published elsewhere.

If, now, we summarize the heads of the above remarks in the shortest form, it will be seen that the fungi and their allies[1] which concern us here may be put into a list as follows—

1. The *Peronosporeæ*, and their numerous allies.
2. The *Ustilagineæ*, and their related forms.
3. The *Ascomycetes*, containing at least four large sub-types.
4. The *Uredineæ*, or Rust-Fungi.
5. The *Basidiomycetes* (Mushrooms, Toad-stools, &c.).

Now it is a curious fact, that in all these groups we find different stages or degrees of parasitism, and in various cases it is possible to put our hands on three kinds of forms, (1) such as *must* obtain their food from a living host, and are therefore strict parasites; (2) such as *can*, and usually do live for some time on dead organic remains, though

[1] The *Myxomycetes* are not fungi properly speaking; nor are *Bacteria* fungi.

they occasionally attack living hosts and live parasitically; and (3) forms which are incapable of obtaining their food from other living organisms, but *always* live as saprophytes.

Several points of great importance follow from what we already know of these fungi. In the first place, I could show you certain green Algæ, which are able to manufacture their own carbon compounds because they contain chlorophyll, but which so strongly resemble the fungi of the *Peronospora* group in all other essential respects, that even the dullest comprehension must apprehend the position that if one of these Algæ, from any cause whatever, ceased to develop chlorophyll it would be classed with the peronospora-like fungi. When, further, it became known that many Algæ tend to become parasitic on or in other living green plants, the idea was obviously suggested that probably such fungi as those of the *Peronospora* group are simply the descendants of Algæ, which have become parasitic, and therefore (because they no longer needed the power of forming their carbon compounds from the inorganic world) ceased to produce chlorophyll. It is worth noticing how fruitful this idea has been. It stimulated investigators on all sides to examine every trace of evidence for and against the hypothesis, with the result that we are convinced that the fungi are derived from the Algæ. This is one of the points referred to above.

Another is that the evidence tends to show that the true fungi have all descended from a common pythium or peronospora-like ancestor. As we pass along the lines of relationship between the various forms, it is hardly possible to doubt that the ancestors of the large *Peronospora* group, sending out progeny in several directions, gave rise to forms like *Mucor*, *Phytophthora*, and *Podosphæra*, and perhaps to the *Ustilagineæ* and their allies: the rest of the *Ascomycetes*, at last reaching forms like *Claviceps* and *Peziza*, are probably derived from the ancestors or relations of *Podosphæra*, and some of these lead us to the *Uredineæ* and *Basidiomycetes*. It is, of course, impossible to bring forward in detail here the evidence on which such conclusions rest, and my only object in mentioning these facts is to show the reader what problems are being discussed as our knowledge advances.

Another point is, that in each of these groups certain forms have become strictly parasitic, others strictly saprophytic in their habits; while in others the mode of life is of such a kind that the fungus can take advantage of occasional opportunities for being either the one or the other. In some cases, as in that of the *Peziza* of the lily-disease and its congeners, we have direct proof of these convenient adaptations, which are of enormous advantage to the fungus.

With these remarks I must conclude, as it is beyond my purpose to follow up the numerous other phenomena of life presented by these curious organisms.